NEXT GEN DEVOPS
Creating The DevOps Organisation

Grant Smith

With thanks to my girlfriend Kylie for her constant support and my editor Kate O'Flaherty for turning my incoherent rants into cogent arguments. Thank you too to my Mom and Dad for over 40 years of encouragement and support.

ISBN-13: 978-1517021573
ISBN-10: 151702157X

About the author

Grant has created and led high performance Operations teams in some of the largest and fastest growing companies in the UK over the last 18 years and has been at the forefront of the DevOps movement for the last 5 years. He's driven real collaboration between Operations and Development teams in AOL, Electronic Arts and British Gas by implementing Infrastructure as code and driving application integration from continuous build systems. Grant has delivered game platforms running in the cloud enjoyed by millions of players per day, websites serving a billion page views per month and Europe's largest Internet Of Things platform. In addition to writing Next Gen DevOps: Creating the DevOps Organisation he has also published his own open source DevOps Transformation Framework available on Github. Grant is frequently sought out for his cloud and DevOps expertise and can be reached at grant@nextgendevops.com.

CONTENTS

CHAPTER ONE

Surely we can do better than this?

In this chapter…

We'll see how, in traditionally structured IT organisations, the company leadership and each of the major functions within IT have different and often conflicting demands of the operations function. We will examine the most common compromises operations teams make in order to try and balance these different requirements and expose the danger of these compromises.

Finally we will consider a way of looking at the operations function that resolves these conflicts and includes the technology organisation as a whole in decisions about operational priorities.

Conflicting IT: the hand-off culture

Most IT organisations are structured around the teams that comprise them. The development team develops products, the testing team tests the application to intercept problems prior to launch, and the operations team runs the live service, development systems and ancillary services. This is called a functional organisation structure.

These teams are all overseen by a director or CTO. If more capabilities are needed, teams are added: security, network and first-line support functions can often be found in the larger IT departments.

Requirements typically come into the development team first, which in turn hands off a product to the testing team - which then hands off a tested product to operations for hosting and live support. During my time at AOL it wasn't uncommon for projects to have spent some months in development before the operations team even came to hear of them. This hand-off culture and the way such hand offs are managed sets up a number of dangerous conflicts in the IT organisation.

Operations teams are always asking themselves: Is now a good time to investigate operating system patches or find out why the development team is having problems with the build server? Should the team prioritise making sure that back ups are working properly or work with the development team to investigate product performance problems? What is the relative importance of resolving tickets raised by developers and testers, compared with the need to ensure infrastructure is being used efficiently?

There have, so far, only been half-hearted attempts to mitigate these conflicts. Project managers and scrum masters will try to create friendly working relationships with the testing and operations teams, in the hope that their projects get prioritised higher than other work. Occasionally, operations engineers are invited to project progress meetings so they can pick up on schedule and functional changes. This is a vain attempt by the project managers to try and improve the odds that operations engineers will be available when their project needs them. These measures are never particularly effective..

IT and process: awkward bed-fellows

Senior IT leaders are judged as successful in their organisations when their teams release new products and new features often. The most successful achieve this while generating positive publicity for their organisations and reducing costs.

The IT organisation, the business itself, and the environment in which it trades seems like a chaotic system to the leadership team. When a business is not releasing new products and features often enough, process seems like a good candidate to improve the situation: the more variables that can be turned into constants, the less chaotic the system seems.

Yet process is awkward for the IT industry. There is a constant fear that applying process inevitably leads to bureaucracy and red tape. Unfortunately every senior IT leader has had bad experiences with large centralised IT functions. As a result, they fear that implementing process inevitably leads to a risk-averse culture incapable of flexibility and innovation.

At EA the centralisation of IT and operations functions had become like a campfire horror story. Ooh, beware of processes or you'll end up just like them!

Almost all firms are unhappy with how their IT organisations bring new capabilities to market. Development is always late; test engineers miss too many bugs; operations are blocking the whole process. The build systems are too slow; too much time is spent planning; the hosting is inadequate and too expensive. The list goes on and on.

The development team's activities are managed by a project manager or scrum master who is adamant that the team is focused and working as hard as it can. The test team have become masters of time-management, ensuring people are always available the very second it's possible to execute tests. The operations team sits at the end of the process and always appears wrong-footed and unready when called upon to participate. It seems obvious to everyone that the problem must lie here.

What is often forgotten is that the operations team has been given an organisation wide mandate; operations aren't tasked with just working on just one product or one release. The operations team is trying to balance its time between ensuring it is providing a secure, reliable and efficient estate, keeping the developers and test engineers working, and ensuring the live products are functioning well.

Then, when a project team needs them the operations team has to drop whatever it is doing and re-engage with the project. Where it needs to assess whether changes are required to the monitoring and back up solutions. It has to find out whether new infrastructure is required, and if the product uses new technology that will need to be supported. Is it any wonder that operations appears unprepared, given the amount of context switching forced upon it? And this is all assuming there are no high severity incidents, post-mortem reports or audits underway.

The development team: destined to fail?

The development team is judged on how quickly it can create good quality applications. Every day, someone – whether the CEO, CTO or a product manager – will ask when a new build will be ready. This creates increasing pressure, building an environment where developers feel the entire organisation is waiting on them to finish.

The development team's usual role is to build software to a deadline. If the organisation follows a waterfall project management method the deadline often suffers as new and changing requirements are absorbed by the project. Some organisations choose Scrum and Agile software development methods to allow them to better respond to changing requirements and in this case new requirements can be bartered against old ones to allow the team to hit the deadline.

And when those deadlines are imposed on the development team, it naturally feels under siege – and fears being perceived as inefficient or incompetent. Every project or sprint is an opportunity for the development team to fail.

So the development process is one of constant activity: individual developers are designing and coding as well as building unit, functional and performance tests – and discussing alternative solutions and tricky problems with each other. Each day, there are status meetings, where they will comment on stories or use cases, estimating and re-estimating. On top of this, developers are considering the impact of requirement changes, preparing builds, optimising the build process, testing builds, as well as attempting to replicate bug reports and responding to them.

When all of this feverish activity has finally culminated in a functioning build, it either goes on to a formal testing process, or it goes live. Testing teams usually work in parallel with development teams, and as capability is made available, it is tested. Bug reports are provided, tested and fixed prior to the build being functionally complete, creating a fever pitch of activity between the development and test teams.

The next disconnect

Then emerges the disconnect: having worked solidly for several weeks or months on the project, the development team must sit and wait while another entirely separate team, the operations team - which has had no prior involvement in the project - brings itself up to speed and prepares to make the build live.

There follows a frustrating period for the development team, as operations engineers read the release notes and ask seemingly unimportant questions about implementation details and project decisions that were taken weeks ago.

After spending every working day building, testing and optimising under constant pressure from everyone in the organisation, development has to wait for someone to just change some configuration files and copy some files around. It's therefore no surprise that development gets frustrated. To the development team, every delay is a blocker.

Development simply wants the operations team to deliver what is needed and get out of the way, so it can move onto the next set of requirements.

Testing: 'a necessary evil'?

Testing is seen by some as a necessary evil; by others as an essential safety net; and by a rare few as one of the key factors in achieving real productivity and efficiency in IT.

Regardless of its position, the testing team is always caught between development and launch. Whether waterfall or agile project management methods are used, the time available for testing is always less than planned for: testing leaders often spend as much, or more, of their day scheduling than they do improving the quality of the testing.

The solution to this is usually considered to be a transition of manual to automated testing. This is a great idea, unfortunately the decision to build an automated test suite is is rarely undertaken in a calm climate of requirements analysis and investment. More usually, automated testing projects are initiated in frustration, with teams instructed to do what they can with at best a couple of extra people.

And when test teams try to implement automated testing without sufficient investment, it tends to cause a division within the team. The more technical test engineers find an outlet for their skills and frustrations and are able to create some rudimentary automated systems but then can not be spared for the manual testing that's still the team's primary function.

Meanwhile, the less technical become the only people available for the manual tests that are still required. This sees the test manager's job get even harder: there are fewer staff available for manual testing and now they have to drive a development programme as well.

When communication breaks down: the blame culture

The testing teams are under a great deal of pressure. They need to develop and continue to progress automated solutions. On top of this, they must ensure testers are available as soon as the build is ready to be tested. They also want to verify bugs reported by consumers and raise those with the development team.

While the test team are doing all this, the development team are being questioned by product management and leadership about how all these bugs are making it through to testing in the first place.

The development teams often have trouble reproducing the bugs in their environments and so naturally, the focus shifts to the environments. Inevitably, there are differences between the build, test and live environments: at the very least they are likely to be on different networks or subject to different firewall rules or security groups.

Businesses can rarely afford to run the same scale equipment in testing as they do in live, so compromises are made. These scaling compromises may be functionally insignificant; more often than not, the testing environments are perfectly adequate. If the developers and testers weren't involved in designing them and received little training about them they will be an unknown and therefore a little scary.

Often, in a laudable attempt to improve the efficiency of the build, testing and release process, the development and testing teams begin evolving their processes. This leads to changes in how the environments are used. Suddenly, what was intended to be a build environment used by development for unit testing is being used for integration.

A lack of knowledge of how the environments are configured leads the development and test teams to blame them when bugs make it through to production. The operations team is now a blocker to the testing of the new release because the environments aren't good enough.

This was the first problem I was called upon to fix when I joined Playfish. The games teams were convinced that the development environments were completely different to the live environments and were blaming those differences for bugs found after releases. A little investigation proved there were no significant differences. The problem was education. Playfish had grown quickly, the original developers had a hand in designing and building the environments but the later hires hadn't and no documentation was available. I provided a little diagram and some notes about configuration files and properties on a wiki page and the focus quickly shifted back away from the environments.

Understanding operations

The operations discipline isn't well understood, even within the field itself. It's a discipline that has changed constantly and provides prospective operations engineers with few opportunities for formal training.

Operations engineers tend to be self-taught and find their way into positions by internal movement. Operations engineers are also a rare breed: in a company with 100 developers and 20 testers, there might only be three or four. With little understanding of the discipline, and limited exposure to experienced operations engineers, it's no wonder that so few employers are able to accurately articulate what they really want from their operations teams. It's rare to find operations teams with confident, strategic leadership who can brief new employees on what's expected of them.

On joining an organisation, operations engineers are usually given some rudimentary reading material and asked to work on the ticket queue, or to develop a small application.

Having achieved some reasonable degree of independence on the ticket queue - or built a reasonably functional application - they will typically be asked to undertake a deployment. This is considered their baptism of fire: if successful, it usually culminates in the engineer being considered a fully-fledged independent member of the operations team, needing little further formal supervision. This induction creates a competent, if insular operations team capable of reacting to most live service or environment problems.

Receiving no further input, the operations team does it’s best to build environments that meet the needs of the development and test teams. It will try to ensure irreplaceable data is backed up in as robust a manner as possible given the resources available. It will endeavour to ensure the environments operate in a reasonably secure network and are built with tried and tested versions of the operating system and applications. It does all this because as far as they can tell it's what's expected.

The 'them and us' mentality

On top of their various other tasks, operations is expected to monitor the live service and respond to any incidents. Whether the fault lies within the network, server, off-the-shelf application or in-house developed product, it's assumed the operations team are responsible for investigating and troubleshooting. Often it's assumed that they'll be on-call 24/7 as an implicit part of their duties.

This is where all the previous conflicts start to come to a head. The organisation assumes that operations are responsible for identifying and troubleshooting every incident. If the fault appears to be part of the in-house developed product, the operations team is expected to prove it beyond reasonable doubt before engaging with the development team. After all every development hour lost to troubleshooting impacts revenue. However it also makes the operations team feel accountable for choices it had no hand in making.

Operations engineers tend to love a challenge and immediately start trying to prevent problems from occurring in the first place. In theory, this is exactly what we'd like to see everyone in the organisation doing. In practice, the leadership wants that release deployed now and another conflict is born.

The operations team assumes it's their responsibility to prevent problems occurring and keep systems available. So they will try to prevent system changes being made on Friday afternoons or during the busiest times. They want to perform soak testing before changing major components. They won't reconfigure servers without checking who else might need them regardless of the fact that this might be blocking the new release. This puts them in direct conflict with the developers, testers and pretty soon the leadership.

Operations engineers, developers and testers rarely develop good conflict resolution skills and so the situations tend to get emotional and require escalation. More often than not, the release is deemed essential by leadership - and the changes need to be made immediately regardless of the possible consequences.

And so leadership has to order the operations team to make the changes and reconfigure the servers. The more these situations occur, the more besieged and aggrieved the operations team feel. This is how the 'them and us' mentality develops.

In reality, no-one wants system problems over the weekend, or servers reconfigured without checking with the users first. However, the timing of a single release can be critical. Sometimes a risk has to be taken because the cost of missing an opportunity is too great but understanding this requires context.

Operations on the defensive

The operations team feels responsible for the availability and capability of the service, they are often held accountable for it. Yet they are also often asked to make configuration changes or allow other teams to do so without being given the time to understand the full ramifications of the decision.

When the operations team receives an urgent request to reconfigure systems and is not given sufficient time to investigate the implications, They perceive it as a risk to the service's availability. Service availability isn't assessed over a day or a week, it's assessed over a calendar month or rolling 30-day period.

It's therefore no surprise that the operations team is wary. Its first thought is the long-term availability of the service and the reliable configuration of the environment, rather than the timely deployment of a single release.

The operations team know that at the end of the month, when they are being held to account for availability shortfalls, no-one will care that they were pressured to make that one ill-considered configuration change that caused the outages. It requires time to understand the implications of configuration changes, then to code solutions to include the changes in future builds - and roll them back if necessary. When confronted with this situation regularly, the team starts to behave defensively.

There can be no greater crime against successful teamwork than seeing every opportunity as an opportunity to fail. Nevertheless, these opportunities to fail are an inherent feature of the traditional IT organisation structure. Each team supposedly has the same overall goal, but they actually have very different daily concerns.

If the infrastructure is reasonably stable and robust, the leadership often sees no reason to invest any further in operations. They will turn their focus on the most expensive activity in the organisation: building, testing and deploying the product. This forces the operations team to make compromises on all their less obvious activities.

Least investment, biggest cost

Security, data management and configuration management traditionally receive the least investment and focus yet they pose the greatest risk to growth and profitability.

All online enterprises are exposed to a moderate amount of security risk: just having a system connected to the internet is a danger. But having a successful business that processes millions of transactions online vastly increases the size of the target. Add a successful marketing and PR strategy into the mix and your systems become a pulsing, red bull's-eye, attracting every script kiddie, botnet and capable hacker.

Security considerations are rarely the focus of attention for a new company. After all, there isn't much need for more than a good set of firewall rules, some basic OS hardening and sensible network design. But as the business grows, so must security mechanisms' sophistication.

Build and test systems become more sophisticated as products grow: they need to process realistic-looking data. Developers may need to run tests against production systems, and verify in the production environment. Services performing multiple functions will be added to the mix and updated with new capability.

This can lead to problems as the initial basic firewall rules are appended to and changed without an over-riding guiding policy. While all this is going on, operating systems fall whole version numbers behind. Network segmentation is eroded by the need for increasingly sophisticated integration testing, fundamentally compromising the original security vision.

Understandably, the operations team is concerned about this. But at the same time, it is striving to ensure that nothing stands in the way of the build, test and deployment process in case they incur the wrath of the whole business. This is only natural after-all; people react much more strongly to clear and present danger than they do to distant potential problems.

Configuration management: a complex solution

Configuration management brings similar issues. On the one hand, a new business needs little more than a simple OS deployment mechanism, hardening scripts and perhaps some defining of variables to take into account the different roles the systems will play.

However, these systems rapidly become a hindrance. As the business grows, more scalability, sophistication and flexibility is required. The pressure on the operations team to reconfigure systems manually increases exponentially with the growth of the development and test teams.

And suddenly, not having an extremely sophisticated and scalable configuration management system is a huge disadvantage. But who should have developed it and when? Open source configuration tools such as 'Puppet' and 'Chef' provide the sophistication required, but they don't work out-of-the-box. Therefore, bringing these tools into an enterprise that already has existing standards often requires significant development time.

Newer firms are lucky in this respect: operations teams choose tools such as these and development can progress over time. Yet for a business that started five years ago and needs sophisticated and scalable configuration management, the level of effort required is nothing short of Herculean.

This tale repeats for backups: when a business is small, taking a database dump and duplicating the file into a mass-storage device is a perfectly adequate backup mechanism. More sophisticated mechanisms are required for larger firms experiencing greater rates of change against more products.

However, the people who need to grow and flex configuration management and security - as well as keep pace with the development and test capability - are also the ones who should be assessing backup mechanisms and strategies, and testing restores. They are also the ones called upon to help the development and test teams and reconfigure systems and perform the deployments and respond to all the incidents and assess the capacity requirements of the organisation.

Operations as a product team!

Analysed over the short term, the requirements presented to the operations team by development, testing and leadership directly conflict with each other. However with a little change in perspective, these requirements can align perfectly. With planning and investment, most of these requirements can be delivered with little effort on the part of the operations team.

Rather than looking at operations as a service function, it should be thought of as a product team.

The products chosen, built, maintained and supported by the operations team are:

1. Configuration management
2. Continuous integration
3. Monitoring and dashboards
4. System integration
5. Data management
6. Ticketing and documentation

By considering these functions as products in their own right, it becomes evident that each needs a set of defined requirements agreed with all stakeholders and roadmaps describing their future development.

Each product requires an implementation budget and commercial-off-the-shelf (COTS) tools must be assessed against the resource requirements for building bespoke solutions. The efficacy of each product is determined by judging its performance against agreed metrics.

Just as the leadership needs a mechanism for determining which features the development team builds and which bugs are prioritised, it requires a mechanism for agreeing how resource is allocated to choosing, building, maintaining and supporting these products.

By considering this as operations' core focus, asking how many engineers are needed per server, or per developer, is unnecessary. Instead, the IT organisation can examine the service it builds and determine how sophisticated each of the operations products should be in turn.

Let's take a look at each and see how they integrate as individual products, resolving conflicting requirements coming in to the operations team.

1. Configuration management

Configuration management is a familiar concept. This product is concerned with how servers are built; but it doesn't end there.

Configuration management centres around every aspect of the server's function and scale. Specific configuration changes are needed for different applications, and the product must understand how the system should be built or re-configured depending on these variables.

For example, a system used for development might utilise a collapsed stack running web server software alongside Java Virtual Machines (JVMs), application servers and database software solutions. On the other hand, if it is intended to support products in live, some of these applications might be split onto dedicated systems.

When systems are used for integration testing, firms often run multiple solutions alongside each other so the billing service, content management solution and customer relationship management (CRM) might all share hardware.

The configuration management product must have the ability to build, re-build - or re-configure systems - as requested. If these are common requests more resources can be invested to increase its flexibility, endowing it with self-service capabilities.

If the Configuration Management product is combined with real-time dashboards and system integration products, together they resolve several conflicting requirements immediately. The operations team is no longer concerned about surprise changes: dashboards and systems will be updated as configuration management products make adjustments at the request of development and test teams.

In situations where reliability is more important, configuration management can be augmented with features such as change gates and notifications giving leadership direct control over the changes.

2. Continuous integration

Continuous integration and deployment involves building services, from source through development, test, integration and live environments, in an automated and repeatable fashion.

But all too often, this process is stilted. Development teams progress a product through build environments into testing themselves, packaging it so operations can deploy it into live. This is time consuming; it exposes the process to unnecessary human error; and also introduces an element of mistrust. The packaging activity itself is fraught with risk and complexity as it's highly dependent upon the environment where the package is created.

When there's a problem following a deployment, the first question asked is whether it was performed correctly. There is no need for this process to be manual: deployments involve building the artefacts, copying them and making a few configuration changes.

Imagine a scenario where the same tool prepares and deploys into development, build, integration, test and live environments. Picture that this process is controlled and managed by the configuration management product and that each stage updates a dashboard and tickets.

It offers the control and visibility that was present before, but the opportunity for human error has been almost entirely removed, with minimal delay between deployments. Self-service combined with appropriate environment rules defined in the configuration management product ensures no environment configuration problem ever blocks development or test activities.

3. Monitoring and real-time dashboards

Monitoring and dashboards have been ranked third, but it's this capability that should be available first. In fact, it's monitoring and dashboards that really promote and enable the benefits of configuration management and continuous integration.

Monitoring centres around running regular tests against the service and reporting the results. Alerting involves analysing the results of tests run by monitoring, letting the organisation know when they match pre-defined conditions.

Meanwhile, dashboards display the results of tests and other real-time data to inform the organisation about the things it cares about. And these three functions form the monitoring and real-time dashboards product.

In other words monitoring systems are automated test suites. Hence tests run against the service should be designed in concert with the development, test and leadership teams, with definitions, metrics and thresholds agreed upon. If the tests are designed with the development team, it is possible to create bespoke monitors that actually report on the state of the service itself - rather than having to infer this from several infrastructure tests that only provide circumstantial evidence at best.

If the thresholds for these tests have been agreed, no debate is necessary. It's also possible to work out action plans in advance: hardware can be thrown at a problem as an agreed temporary measure until more permanent solutions are found. In addition, dashboards can be enhanced with a wealth of real-time data easily available to the operations team.

4. System integration

It's rare these days to find an online service that stands alone: more commonly, they depend on other solutions, whether in-house or third party. Once all the undocumented features are understood, configuring these services to work together is fairly trivial in production because the values rarely change.

But in the development and test environments, this is far from trivial: the services need to work differently for any given test. It's easy to lose track of the changes made, leaving systems in a poor state.

The goal of a system integration product is to map relationships between different configuration options and allow them to be changed and modified as necessary, while tracking changes. Sophisticated systems will allow preset options to be saved, so that configurations can be switched quickly. All this capability can be made available through self-service interfaces, allowing the development and test teams to make necessary adjustments without fearing that systems will be left in a bad state.

5. Data management

Many organisations are backing up Gigabytes of useless data because policies that were set 10 years ago haven't been reassessed.

Newer organisations are backing up data that they would never restore simply because that would take longer than building systems from source.

The 'Millennium Bug' scare in 2000 led many companies to invest in disaster recovery (DR) or business continuity (BC) plans. This allowed operations teams to find executive sponsors and get backups in good order. Yet DR and BC have fallen out of fashion again.

Almost without exception, operations teams know every failing of their policies, tools and processes, but feel they are unable to do anything about it. They just can't spare the time or find an executive sponsor to support the initiative.

However, formally initiating a data management project will highlight internal systems weaknesses. Subject to internal analysis and debate, the resources needed to deliver requirements and capabilities can be assessed honestly. When there is an opportunity to discuss and agree the metrics that will be used to assess the success or failure of the product, appropriate investment can be acquired.

6. Ticketing and documentation

In businesses with multiple products and services installed on multiple systems throughout their estate operations teams can struggle to identify exactly which software artefacts and services are at work on each system.

There was a time in that system's life when all the components were known and well understood. That was at deployment time. However records are rarely kept of the state of systems prior to and just after deployments. Release notes provided by the development team are often all the operations team have to go on.

Now imagine the system identity is linked to data describing the version of all the applications deployed on it. All changes have been made, with direct links to release notes and tickets raised against that system. With an appropriate system tickets and release notes can be available, at a glance to a responding engineer. They would have all the information needed to make judgements and can take action without having to waste time gathering data.

This is just part of the remit of the ticketing and documentation project. It is the part that would benefit the operations team most. However formally defined ticketing and documentation products with recognised owners benefit the entire business - in many different ways.

Conclusion

Let's look again at the conflicting requirements facing the IT organisation:

- The leadership wants predictability and stability but also needs releases live as soon as they are ready.

- The development and testing teams need releases deployed immediately; they need build and test environments to behave exactly like live environments; and they need the flexibility to deploy applications and people where they need them at a moment's notice.

- The operations team feels responsible for the availability and capability of the service and yet it is often asked to make configuration changes or allow other teams to do so with no time to assess the risks and impact of such changes. This forces them to compromise the less visible aspects of their remit such as security, data retention, business continuity and providing the automated solutions their peers need.

If, rather than considering these requirements as functions to be performed by particular teams, you consider these requirements with a product view suddenly they do not conflict.

As the operations product portfolio is built and specified, each requirement is assessed against the others. They can be grouped together and products can be designed that meet the needs of the entire IT organisation. In doing so, operations becomes demystified. Everyone in the company sees the difference between installing Mint Linux on their laptops and operating online products and services, and the operations team immediately becomes a real part of the technology organisation, with everyone having an opportunity to influence its priorities.

This product view can provide many more benefits than just improving the quality of service provided by the operations team. The rest of this book will consider how online products and services are provided by IT organisations and describe the next essential step in the evolution of the IT industry.

CHAPTER TWO

The History of Operations

In this chapter...

We'll look at the history of the IT industry. George Santayana the Spanish/American philosopher once wrote: "Those who cannot remember the past are condemned to repeat it."

Prior to writing this book I knew almost nothing of the history of my profession. I'd read about the first computers and the events that led to their role in the modern day. I'd read about the first programmers and the first programming languages. I'd even read about the history of the major operating systems that still shape how we use computers today. However I had no conception of how the IT industry came to be how it is or how the operations function came to be at all.

In this chapter we'll look for clues to help us understand why modern IT organisations are structured the way they are. In particular, we'll unearth evidence to explain why teams are so often split by skill set and why operations is not typically considered part of the development process.

Human computers

For the 200 years prior to the end of World War II, computers were their era's equivalent of calculators. A mathematician would define an equation that was needed with a range of variables. The computer's supervisors would break this down into simple arithmetic operations, and each would be assigned to a computer.

The combined output from computers was collated, with large numbers of calculations made in this way: timings of astronomical phenomena, pay roll and ballistics were worked out like this for centuries.

This photo was probably taken in the US Treasury Department sometime between 1909 and 1932. Image courtesy of the Library of Congress.[1]

---

[1] http://www.loc.gov/pictures/item/90710989/

But later, the advanced technologies employed in the second world war made this impractical. New weapons were being developed and the volume of ballistics calculations alone was swamping the US's human computing capacity. This was before magnetic mines and the required drop-off calculations, let alone the frighteningly complex spherical implosion calculations needed for the atom bomb.

While, technically at least, the electronic computer revolution began in peacetime, the desperate need for vast quantities of calculations brought on by the second world war made the funding available for computers to change the world.

Electromechanical and electrical computers

The digital computer revolution began in 1936, kicked off by a German civil engineer named Konrad Zuse. Zuse was, quite understandably, frustrated with the calculations required by his profession.[2]

As impressive as his story is, we're not concerned here with the history of computing, but that of operations. To witness its inception, we have to look back to Harvard University in 1944, the home of a computer called the ASCC - or Harvard Mark 1.

The Harvard Mark I was designed and built by Howard Aiken and IBM for the US Navy. Aiken built a remarkable small team to programme and operate the Harvard Mark I: Richard Bloch, Robert Campbell and Grace Hopper are widely acknowledged as the

---

[2] ComputerHistory. (2009, July, 1 recorded 1996). Computer Pioneers - Pioneer Computers Part 1. Retrieved from https://www.youtube.com/watch?v=qundvme1Tik

world's first programmers. They are less well-known as the world's first operations engineers.

The Harvard Mark I's wartime team, early 1945.[3]

Once a programme was running, the responsibility for its smooth operation was theirs: they were on-call 24/7 to provide expert support. Hopper often commented that she and her fellow

---

[3] Image reprinted with the kind permission of the Collection of Historical Scientific Instruments.

Officers on the second row: Ensign Richard Bloch, Lieutenant Commander Hubert Andrew Arnold, Commander Howard Aiken, Lieutenant Grace Hopper, Ensign Robert Campbell. The Harvard civilians in the back are technician Robert Hawkins, secretary Ruth Knowlton and Assistant Director of the Cruft Laboratory, David Wheatland.

The enlisted men are Seamen Livingston, Bissell, Calvin, White, and Verdonck.

programmers knew the machine better than those that built it and as a result, they were often able to direct engineers straight to the actual failed relay.

The next milestone on the journey to discover the early operations engineers leads to Philadelphia in 1946, where John Mauchly and Presper Eckert built ENIAC (Electronic Numerical Integrator And Computer). ENIAC didn't store its programmes in code: while a general purpose electrical computer, the ENIAC was programmed electro-mechanically, with switches set and plugs changed to store the data.

J. Presper Eckert and John Mauchly.

Just like the Mark 1, programming the ENIAC involved breaking down equations into simpler mathematical operations. The programmers had to assign where values would be stored for later computation and ensure that numbers were recorded on punch cards when they became too large for storage in the computer's memory. This was the job of a group of women who became known as the 'ENIAC girls'.

Three teams of women programmed and operated ENIAC in the 1940s and 50s: Kathleen McNulty, Antonelli Mauchly, Jean Jennings Bartik, Frances Synder Holber, Marlyn Wescoff Meltzer, Frances Bilas Spence and Ruth Lichterman Teitelbaum.

The ENIAC girls were programmers first and foremost: their role was to take mathematical equations and reduce them to a programme ENIAC could operate. They weren't electrical engineers and certainly didn't design the circuits or any of the arithmetic units. However, when ENIAC failed - and at first this happened often - these people investigated, calling hardware engineers when components needed to be replaced.

Marlyn Wescoff (standing) and Ruth Lichterman wiring the right side of the ENIAC with a new program, in the "pre- von Neumann" days.[4]

Betty Jean Jennings (Bartik) once commented:

"The biggest advantage of learning the ENIAC from the diagrams was that we began to understand what it could and what it could not do. As a result we could diagnose troubles almost down to the individual vacuum tube. Since we knew both the application and the machine, we learned to diagnose troubles as well as, if not better than, the engineer.[5]"

---

[4] "U.S. Army Photo" from the archives of the ARL Technical Library.

[5] Barkley Fritz, W. (1996, September). The Women of ENIAC published in IEEE Annals of the History of Computing Volume 18 Issue 3, September 1996 Page 13-28

As well as troubleshooting faults, these early programmers/operators extended the capabilities of their computers. Richard Bloch designed a subsidiary sequence mechanism that replaced the need for operators to load new tapes to branch programmes. This mechanism allowed the Mark 1 to handle 10 different sub-sequences effectively, giving it subroutine and branching capabilities.

Adele Goldstein and Jean Bartik developed the ENIAC into a stored-programme computer, with the function tables containing the coded instructions. The very earliest of the computing pioneers were not only programming and operating their machines, they were upgrading and even designing completely new features for them.

Left: Patsy Simmers, holding ENIAC board Next: Gail Taylor, holding EDVAC board Next: Milly Beck, holding ORDVAC board Right: Norma Stec, holding BRLESC-I board.[6]

The birth of the OS

The 1960s saw the transistor replace the vacuum tube, mini-computers join the mainframes and the explosion of the computer industry. Two events during this era had the largest influence on the operators of the time: the first was the explosion of demand for computers and operators to work on them; the second was the development of the operating system (OS).

Encouraged by President Kennedy's tax cuts, US businesses were growing at an incredible pace. Hungry for an edge, many large firms saw computers as a way of increasing their speed and efficiency. These early computers were primarily used for data analysis and payroll.

During the early part of the 1960s, most companies bought their computers complete with the programmes needed. All the major computer suppliers of the day provided application programming services. This meant that while businesses investing in computers didn't require their own programmers, they did need to hire computer operators to manage and maintain machines, organise data and collate results.

---

[6] U.S. Army Photo, number 163-12-62

IBM type 704 electronic data processing machine used for aeronautical research.[7]

The computers of the 1960s were mainframes: they had intelligent input and output systems capable of independent operation from the main central processing unit (CPU). Data transfers were handled by channels controlled by their own mini-computers. Once the data had been located and transfer begun, the CPU was free for more work.

Huge print jobs could operate for days independent of the CPU. The huge tape reel machines that appeared in 1970s TV series 'The Six Million Dollar Man' weren't just fancy tape recorders: they were computers in their own right, designed to take work away from the main CPU allowing it to return to controlling and executing programmes. The management and maintenance of the main computer and ancillary subsystems was the remit of the computer operators.

---

[7] Image courtesy of NASA Langley Research Center.

These large, complex mainframes containing semi-autonomous subsystems and vast quantities of magnetic tapes required large numbers of people to operate and maintain them. Therefore, as the decade progressed, there was a growing feeling that computers weren't justifying the investment. It was widely felt that the machines should to do more work, without needing more people to manage them.

In addition, those organisations with computers were suffering serious resource constraints as an increasing number of uses were found for the machines. Unfortunately, computers in the 1960s could only execute one programme at a time and if this failed or required input, the machine would sit idle awaiting instructions.

Enter the Operating System

The first operating systems were simply batch processing applications. They removed the need for operators to manually pull tapes from job queues by allowing programmes to be lined up by the computer itself.

Created in 1956 by Robert L. Patrick of General Motors Research and North American Aviation's Owen Mock, the first of these early operating systems was GM-NAA I/O (General Motors and North American Aviation Input/Output).

It wasn't until the development of the Compatible Time-Sharing System (CTSS) in 1961 at Massachusetts Institute of Technology (MIT) that time-sharing was introduced. This brought in the ability to schedule parts of programmes to run during idle periods, making computers much more efficient. Prior to that, if a programme failed and was being debugged or required input, a machine could only wait.

Fernando Corbató and Robert Fano, who worked on CTSS, wrote:[8]

"For professional programmers, the time-sharing system has come to mean a great deal more than mere ease of access to the computer. Provided with the opportunity to run a program in continuous dialogue with the machine, editing, 'debugging' and modifying the program as they proceed, they have gained immeasurably in the ability to experiment. They can readily investigate new programming techniques and new approaches to problems."

[8] *Hauben, M, Hauben, R. (1997 April). Netizens: On the History and Impact of Usenet and the Internet. Wiley*

Robert Fano and Fernando J. Corbató at the Multics Reunion and Dinner 2004[9]

This one significant development revolutionised the role of the programmer and computer operator. Together, batch processing and time-sharing automated much of the clerical work associated with computer operation.

The hardware was changing as fast as the software. The first generation of computers were electro-mechanical behemoths with workstations akin to mission control stations from 1950s Saturday Matinee features. By 1961, the Univac III looked like an electric typewriter: complexity had moved on from electro-mechanical to electrical and software.

But this wasn't readily apparent to the budget holders. Eugene F. Klausman of Remington Rand Univac realised this, using it as ammunition to present a paper entitled: 'Training the Computer Operator' to the 16th Annual Meeting of the Association of Computing Machinery. He wrote:

"The operator's responsibilities include the running of production programs, programs being 'debugged', and service routines, such as compilers, tape correction routines, etc. The operator's responsibility also includes the diagnosis and action taken as a consequence of transient errors.

"In addition is the general area of communications into which the operator fits. To intelligently operate the system his knowledge should transcend mere ability to push buttons, an activity which may steadily decrease with the growth in sophistication of the programming art and engineering..."

---

[9] image reprinted with the kind permission of Justin Knight photography, http://justinknightphoto.com/

Photo copyright Larry Luckham, Reproduced here by permission.[10]

While not entirely bespoke, computers in the 1960s were specifically configured for a particular function. Operating system configuration and application programming sat alongside the hardware arrangement, making investing in new initiatives or business change very expensive: computers would need to be reconfigured by the vendor.

[10] http://www.luckham.org/LHL.Bell%20Labs%20Days.html

Photo copyright Larry Luckham, Reproduced here by permission.[11]

IT giant IBM sought to increase returns for companies investing in computers by giving them the ability to configure machines themselves as their business and markets changed. This led to the launch of the IBM/360 and its System/360 OS.

In order to unlock the power of System/360, companies needed a new role: the system programmer. These programmers were able to provide specific custom configuration of the OS itself as well as the applications running on it. They also created new programmes to streamline computer operation and take new business processes into account.

[11] http://www.luckham.org/LHL.Bell%20Labs%20Days.html

Interactivity

A decade on, the world of computing was still focused on efficiency. But something else had the industry excited: interactivity. ATMs were introduced and quickly became commonplace.

Anyone working in the travel industry will be familiar with Sabre (Semi-automated Business Research Environment), which like the ATM is a classic example of computer/human interaction from the 1970s. Originally developed by American Airlines in the 1950s to handle reservations, it was later migrated to IBM's System/360 and by the mid-70s, was used by travel agents directly.

The need for more flexible interactivity contributed directly to the development of the minicomputer. Whereas the cheapest mainframes were costing hundreds of thousands of dollars, minicomputers' simple design meant they could be purchased for around a fifth of the price.

It's no surprise that they soon found their way into smaller businesses, the manufacturing industry, and most importantly, universities. The minicomputer market came to be dominated by DEC, whose PDP range sold in the thousands - especially the PDP-8 devices which came to prominence during 1960s.

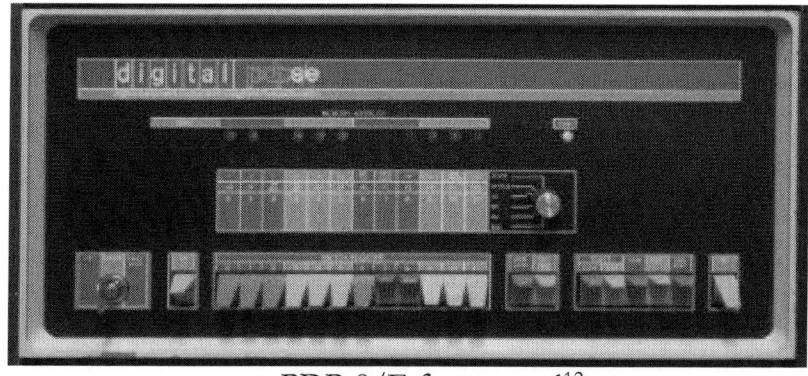

PDP-8/E front panel[12]

While it went through many iterations and was superseded several times, the popularity of the PDP-8 and the applications available for it ensured it kept selling. It's estimated that by the time the PDP-8 ceased production in 1979, around 300,000 had been sold.

Enter Unix

Day-to-day operation of minicomputers had little in common with mainframe operations. While mainframes can be categorised by their channels and intelligent input and output devices, minicomputers saw the collapse of these.

Specifying a minicomputer was similar to a mainframe: there was a huge catalogue of modules available, particularly for the DEC PDP range. However, the minicomputer was a single machine and once delivered on-site, it just needed to be plugged in. It was even portable.

But the minicomputer was much more significant than this: it paved the way for the next great innovation in computing, Unix.

By the 1970s, the importance of the operating system was undisputed. As all major computer manufacturers knew, the OS made the most of their hardware - and it was universally acknowledged that a computer's performance was relative to its operating system.

Computers had already found their way into many large enterprises, with newer minicomputers bringing technology to a

---

[12] photograph by Florian Schäffer. Image reduced to 1/10th size. Reprinted here thanks to the CC BY-SA 4.0 license: http://creativecommons.org/licenses/by-sa/4.0

wider pool of firms. But inevitably this also highlighted the problems inherent in those computers, and so grew a desire for what became known as 'utility computing'.

Utility computing allows many different people to use - and be charged for using - resources while providing discrete and protected access. It also expresses a desire for reliable and robust computing environments where resilience is achieved through multiple independent systems, rather than complex individual ones.

The inability of the early mainframes to provide utility computing led to MIT, General Electric and Bell Labs to begin work on Multiplexed Information and Computing Service (Multics). Multics was a very ambitious project that aimed to offer telecom-levels of availability coupled with enhanced features such as a hierarchical file-system and role-based access control per file.

But by 1969, Bell Labs decided that the project wasn't going to deliver valuable results in time and suspended its activities. Two Bell Labs scientists working on the Multics project were extremely disappointed. Dennis Ritchie and Ken Thompson had become very used to the advanced computing environment the project offered, even if it could only support a small number of simultaneous users at the time.

Ken Thompson and Dennis Ritchie. Image taken from the Jargon File.[13]

This inspired them to continue with it in another form - which was to eventually become Unix. In October 1973, Ritchie and Thompson presented a paper, 'The UNIX Time-Sharing System', introducing Unix formally to the world.

There are many things that made Unix special, but one overriding factor almost guaranteed its success. In 1949, the US Department of Justice (DoJ) began pursuing an antitrust suit against the owner of Bell Labs, AT&T, alleging it had been involved in price fixing. Eventually, an agreement was reached. The terms included the stipulation that Bell Systems patents be licensed to competitors on request.

AT&T had many patents and these were now to be licensed without royalty payments. In addition, future AT&T patents were to be licensed to applicants at "reasonable" royalty rates, with necessary technical information shared.

[13] http://www.catb.org/~esr/jargon/html/index.html

This agreement made it possible for everyone to make use of and contribute to Unix - particularly universities, many of whom were using machines similar to the PDP-11 that the operating system was originally built for.

However, Unix and AT&T's patent situation alone wouldn't have been enough to guarantee the widespread adoption of Unix. The final key to its success was Dennis Ritchie's C-programming language and the port of Unix to C. Now, for the first time, the world had an operating system with limited licensing implications that could be easily ported to any hardware.

Unix's impact on computer operators can't be overstated: what was a profession dictated and limited by vendor which isolated each operations team within its company now allowed for common practices to be developed and shared.

Universities could now make significant contributions to the profession. Programmers could graduate from higher education with at least some relevant experience in the computer systems that they would be using at work. Computer operators moving jobs would now have a much shallower learning curve and could contribute more, earlier on, with the ability to build further on others' work.

Distributed computing

By the 1980s, networking was nothing new; LANs and WANs were relatively common in large enterprises. However, the development of the microcomputer (PC); the commercial development of packet switching networks and the emergence of an Ethernet open standard paved the way for distributed computing to dominate the IT industry.

PCs controlling cheap microprocessor sensors networked back to mainframes revolutionised heavy industry, allowing it to have process control systems long present on assembly lines. Meanwhile, the computer power and graphics processing available in PCs created new possibilities for human interaction.

The mainframes didn't go away: they are, after all, still with us today. However, the PC made it possible to produce more usable and functional interfaces. These systems were still, functionally, little more than terminal emulators: data access was through file transfer between the desktop and the mainframe.

During this time, operations teams in large enterprises were still focussed on the mainframes and midrange computers. Yet operations roles were now appearing in much smaller businesses.

Growth of the PC

The explosive take-up of workplace PCs was phenomenal, growing from 1.2 million computers in 1981 to nearly 13 million in 1986.[14] But now that very small businesses and individuals could afford computers, there was a need to network them and share data.

Methods of data sharing were evolving, with products such as Novell's Netware making it possible to share individual files between networked machines. This saw the advent of the LAN server.

The operating paradigm was similar to the way in which users interacted with mainframes. However client-server networking saw

---

[14] U.S. Congress, Office of Technology Assessment, (Washington, DC: U.S. Government Printing Office, September 1989). Statistical Needs for a Changing U.S. Economy, Background Paper, OTA-BP-E-58

servers designed with the assumption of significant end-user processing power - rather than assuming the terminals were dumb and the mainframe had to provide all the processing power.

And so the LAN server marked the birth of the hobbyist operations engineer: with relatively little money, a motivated individual could buy a couple of PCs, install Novell Netware or a Unix variant and build a network in their home. This was how I started in the early 1990s.

Prices were coming down dramatically: whereas in the previous decade $20,000 bought a minicomputer that could be shared by a dozen users, in the 1980s it could buy a server and 10 workstations with centralised resources and no competition for processor time.

After IBM developed the PC, Unix workstations began to appear, with companies including Apollo, Sun Microsystems, SGI and Hewlett Packard building the machines. Unlike PCs, which used Intel x86 processors, the workstations were usually powered by the Motorola 68000 series CPUs. Far more powerful than their PC equivalents, they were expensive but much more reliable.

Running Unix meant their networking was again considerably different, from both the mainframe world and the PC client-server model. The use of Unix and the relative power of these workstations made them the go-to computer for CAD, commercial graphics work and for universities as science workstations and replacements for the minicomputers.

The massive growth of distributed computing inevitably invigorated the software industry. The US software industry was worth $2.7bn in 1980 and by 1990 it was a $30bn business.[15] The first big sellers for the enterprise PC market were spreadsheet and word processor applications followed closely by CAD.

---

[15] Software Industry. (n.d.). In Gale Encyclopedia of US History. Retrieved May 28, 2014

The fragmentation of operations

It became increasingly common to find computer systems from multiple manufacturers across an enterprise. Some began to break up their operations teams, with those working on distributed computing split out from those assigned to mainframe or midrange - and desktop support separated from distributed computing. This saw the operations profession become as fragmented as the computing environments it worked with. Splits were often acrimonious as teams and the technologies they supported were pitched against each other. The culture in IT for the next two decades was to be harmed by this.

Efficient use of centralised computing resources was a defining characteristic of the previous two decades. Computing power was expensive and so jealously guarded and carefully managed. One of the most important jobs of the enterprise operations team was managing cross-charging. Distributed computing teams were born from the advent of cheap computing; resources weren't something to be carefully conserved. If more power was needed, additional computers could be bought. Universities began to teach developers not to optimise; their time was expensive compared to the cost of computers.

This threw computer operators' role as guardians of the system into sharp relief. Whereas in the mainframe world, poor quality code could halt businesses for entire days, for a well-designed distributed system, inefficient or even bad code should have a much lower impact.

The Internet

Mainframes, midrange computers and PCs were still prominent fixtures going into the 1990s. When I started on the tech support helpdesk at CompuServe, the call ticketing system was accessed via a terminal emulator on desktop PCs that connected to a midrange computer.

We were also connected to a central Novell Netware server that provided access to shared files, print services and messaging. We used Lotus Notes for the knowledge management system, which ran on a different server. The arrangement - which was reasonably common in the mid-1990s in this case involved four operating systems: Novell Netware, DOS and Windows, OS/2 and Unix.

These systems were managed by different groups. The first, supported by maintenance contracts, managed call routing and ticketing systems. Meanwhile, the IT support group was backed by engineers that specialised in Novell and OS/2 to run the servers, with another set managing the desktop environments. Over at the head office, a different group managed the CompuServe UK website. Meanwhile, in the US another set of teams managed the email and the 36-bit mainframes that were the backbone of authentication systems and walled garden content.

Operations in the '90s

As computer systems diversified, engineers were pressured to specialise. This became the operations story of the 1990s. Job advertisements would specify the exact OS and software mix that needed to be supported. It saw those who stayed generalist told that their options were limited and salaries would be impacted. For some time - particularly as the middleware industry grew - this looked to be true.

In the mid-1990's Windows rose to almost complete dominance. CompuServe entered a company-wide Microsoft deal that saw almost every server environment replaced with Windows NT 4. When I joined the operations team in 1998, the CompuServe UK web portal and all the other sites were running Windows NT 4, IIS3. Siteserver and MS SQL. There was a push to remove Novell Netware from the local network environment, to be replaced by Windows domains.

During this time, home use of the internet was growing - and its impact was huge. Businesses were increasingly connected to the internet, and network security was suddenly essential.

Prior to this, most firms lacked outside connections. Those businesses with WANs or site-to-site connections were able to rely on their point-to-point nature - which ensured their safety. Equally, connectivity to the rest of the enterprise, from these connection points, was limited.

Universities and the military had been working on security for some time, but the area wasn't a common subject within most enterprises. So now that internet connections were essential, new equipment was needed. Routers and switches become commonplace and another field of operations was born.

Network operations became another speciality as the networking industry expanded. Another technology had come to the fore; an additional specialisation was created; and a new discrete team formed. Operations in large US companies in the 90s (I worked for two: CompuServe and AOL) was an exercise in negotiating with independent silos.

Troubleshooting problems needing network diagnosis required the careful marshalling of resources across multiple teams. Network security was separate from peering, which was itself isolated from router configuration. In turn, firewall rules were implemented by a different team, as were the initial configurations in data centres.

Systems connected to LANs have very reliable usage patterns. Even in multinational corporations, people in each office start and end their day at around the same time and the number of users doesn't vary dramatically. In contrast, internet-connected sites can be hit at any time of day, from anywhere: if something becomes popular, usage levels can increase by 10 or even a hundred fold - and this can happen within an hour.

Managing Networks

Internet routes are very unpredictable: yesterday's fast route is today's congested network. If the congested router or switch isn't one of your direct peers and the network isn't adapting quickly, all customers will suffer.

Today's routers have very sophisticated mechanisms and it's rare that a single congested device can have a significant impact on a site. However, in the 1990s it was fairly common for users to experience problems related to a single router on a third party network. We were often left in the situation where, having diagnosed a congested router, we could do nothing but call the company who owned the device and ask them to resolve the problem.

This new paradigm where systems were connected to unreliable networks with unpredictable surges in traffic taxed the system administrators and engineers of the day. New ways of working had to develop: unlike its developer peers, operations as a profession had no history of sharing methods and processes. Online services grew much more slowly as a result.

Increasingly in the 1990s, everyone, at least those employed by technology companies, used a computer running Windows. This created a very antagonistic culture for computer operators - who were now more usually referred to as 'system administrators' or 'system engineers'. There was a feeling that operations required no more skills than the ability to work on a laptop. Firms had little appreciation for the complexity of multi-server systems or high concurrency applications.

While the mainframe computer operators had to put up with being told they were soon to be redundant thanks to PCs, those running x86 servers battled constant confusion as to why they were even necessary given that most computer problems could be solved by just turning it off and on again.

The emergence of Linux

One of the most important developments of the 90s was the adoption of Linux: the operating system provided a real alternative to expensive proprietary solutions and complicated licensing.

Although the growth of the commercial use of Linux can be attributed to many factors, my team adopted it due to the availability of hardware - which we could buy in days, rather than weeks. It was easy to find suppliers who would build the OS, install the applications we wanted and even assign IP addresses to make the machines immediately accessible to the operations team when they were cabled in the rack. This vastly reduced our response time, seeing a host of opportunities open up that might otherwise have been missed.

While businesses were finding out what was possible with the web, they were also learning how to identify and analyse visitors. Meanwhile, e-commerce was taking off, seeing the emergence of several off-the-shelf packages for building and maintaining catalogues and online shops.

Secure processing of transactions directly with banks and payment houses significantly reduced the risks for both businesses and their customers. What started out as a simple way for scientists to share information and peer review papers became an interactive source of news and entertainment, shopping, communication, travel and just about anything else that could be imagined.

Online business

The new millennium brought an increase in the pace of day-to-day life within IT. Previously, when the web was in its infancy, publishing content was a relatively slow affair often involving the operations team.

Ensuring content was uploaded to web servers wasn't a trivial exercise: mechanisms for moving data to multiple machines simultaneously were unreliable. In the Windows world we were using Siteserver, while Rsync was used for Unix. At that time, neither method was entirely reliable. On top of this, checking content propagation was made trickier due to the combination of server, browser and, in the case of AOL, ISP caches.

If everything worked as anticipated, operations engineers would return to developer support, system-related activities such as refining the backup processes; planning and testing the next upgrade; or looking for performance improvements. When content didn't work as expected, the usual round of troubleshooting would begin, with teams inspecting the web server logs, checking database queries and stepping through the code.

XML and the CMS

As internet advertising started to generate significant revenues, companies realised they needed many different types of content to attract people. After all, this would result in more ad views, click-throughs and revenue. AOL, the largest ISP in the UK at the time, invested heavily in editorial staff to create content. The immediacy of the web became an important feature.

The busiest day of the year on www.AOL.co.uk was when the UK government announced new tax rates and duties in its Budget. This would see the site receive at least three times its usual peak traffic.

But as more people became able to view the web at work, every major news announcement started to generate similar traffic levels. Over just a few months, every day became like Budget day. As the need for greater quantities of content increased, it became clear that lovingly crafting each article wasn't going to deliver fast enough to keep drawing visitors back.

Two developments from the late 1990s began to have a real impact on web publishing at this point: the first, XML news feeds; the other, the content management system (CMS).

Even before XML was in common use, the major ISPs and portals had always posted feed data. AOL bought news feeds from the Press Association, Reuters and a few smaller niche organisations. These were originally intended for newspapers and were sold on the understanding that they would be subject to editorial review prior to printing. That wasn't going to work on the web. We had to build increasingly sophisticated mechanisms for weeding out embargoed news stories and posting only those that had already been approved for print. Gradually, news suppliers began adopting NewsML, which allowed us to publish stories directly onto the site almost as soon as they were received.

If ever there was a technology that didn't live up to expectations it was the CMS: I must have been involved in nearly a dozen CMS projects over the years. While there was a blossoming content management industry, neither AOL nor any of the other businesses I've been a part of over the years has ever seemed to find a suitable off-the-shelf solution. As such, we felt it necessary to create our own, this meant we needed to hire skilled application developers. This new in-house capability presented AOL with a wealth of new opportunities.

As the creation of content became a job solely for journalists and editors, the development team were free to create increasingly sophisticated web products. AOL developed competition and lottery systems, as well as fantasy sports and interactive web-based games. The earliest of these games were created with the Cartoon Network's kids channel, but led eventually to a highly-successful online version of 'Who Wants to Be a Millionaire'.

This new interactive content stretched the tools and capabilities of development, test and operations teams across the industry. The caching techniques of the early 2000s made it almost impossible to really push a web server. Interactive content, while still just an exercise in web serving, created a host of new challenges. Connections were held open for longer as larger files were downloaded. Meanwhile, more database connections were required to pull in new information and receive and check inputs.

Three-tier architecture

Over time, the servers hosting AOL's flash games ended up with significantly different configurations to the portal machines. During the late 90s and early part of 2000, servers only really needed to be configured when new software was required. When content became truly interactive, using technologies such as Flash, PHP, Javascript and CGI, each implementation resulted in contrasting usage characteristics and required different configuration.

Around this time, the classic three-tier architecture was born. In a typical setup, front-end web servers would consist of single or dual CPU machines with one or two hard disks and a small amount of memory. Middleware machines might be serving JSP pages or entire applications hosted in application servers such as BEA Weblogic, IBM Websphere, or containers like Tomcat or Jetty.

Typically, these machines were more powerful - often dual or quad CPU with more RAM. At this tier, there might also be different types of web servers, including those allowing databases queries using HTTP. There would be at least a couple of large database servers, which were quad CPU, with huge amounts of RAM and generally large RAID arrays or even network attached storage (NAS). Each of these functions required different OS, firewall and router configuration. It was common to change server configuration every few releases - at least in the first few weeks following the launch of a new product - as bugs were worked out and new usage pattern effects observed.

Of the automated server build and configuration tools available at the time, we used Jumpstart for Sun Systems, Kickstart for Linux and SMS for Windows. With a little investment, these tools were great for creating operating system environments. However, they were much harder to use for frequent reconfiguration.

This led many teams to hand craft server configurations as the requirements and usage changed. The most sophisticated of these teams were creating vast quantities of scripts to allow them to automate and improve the reliability of system reconfiguration. This created even more complexity for operations teams.

Versioning and controlling server configurations had long been discussed but it was far from common. Now operations teams had their own bespoke scripts and configuration files that warranted version control, it was becoming more common to host artefact repositories for system packages, making deployment more reliable. But it wasn't yet common to use these for in-house developed operations code.

One of the great debates at this time centred around how many operations people a company needed. In the late 1990s, the tools for managing servers were almost non-existent, with most tasks requiring custom scripts and bespoke solutions.

AOL's server build systems were very sophisticated, but we still spent at least half of our time creating and adapting tools to manage live production machines. This saw us building stats trending tools to graph SAR data, as well as custom backup and restore solutions for databases and application deployment and configuration tools.

Due to this complexity, even very small setups required large teams by today's standards: it was relatively common for three engineers to handle just a few servers.

Data centre hosting

Over the next five years, the data centre and hosting industry really took off, creating a very visible market for server management software. It was now possible for one engineer to manage 100 servers or more.

However, while the tools used to manage servers improved, the complexity of applications and systems it was a part of increased by an order of magnitude.

This led to tensions between operations teams and those in the technology department - and across the enterprise. On the one hand, leadership wanted technological improvements to reduce the requirement for operations people. In reality increasingly complex demands placed on servers, combined with the immaturity of configuration management tools, created a need for more operations people.

The focus of the operations engineer's role had now shifted from the server, its performance and capacity. It now centred around applications and supporting development teams as they created increasingly sophisticated content experiences.

By now, Linux was well-established as a standard choice for server OS; its capabilities inspired greater confidence in open source products as a whole. Apache very quickly became the default web server; and Tomcat the chosen Java container. While MySQL didn't immediately become the database server of choice, its growth and increasing importance was impossible to ignore: certainly by the mid-2000s it was everywhere.

The impact of open source

Open source software didn't immediately change the lives of operations engineers, but eventually it was to have a profound impact. Previously, software limitations had to be overcome with hardware and architecture. For example, if the development team's chosen application server lacked clustering capability, it was down to the operations team to design network and server configuration to mitigate this. This established a hand-off relationship between development and operations. Product requirements that couldn't be implemented by COTS software configuration and were too expensive to implement in development would fall to operations for infrastructure solutions. A generation of Computer Science graduates had left university with the words: "Hardware is cheap, development time is expensive" ringing in their ears.

Open source software created the possibility of a different type of relationship that has become increasingly common. Now, if a chosen tool is missing a capability, it's quite possible to extend it.

It's also more likely in the present day that someone has already forked the project and extended its capability.

A few years ago at Playfish, we chose Opscode Chef as our configuration management system. Electronic Arts, the parent company, had made a strategic decision to use Rightscale for Amazon Cloud management. We needed Chef to use Rightscale to create instances, rather than call the Amazon API directly.

Chef didn't provide this capability but because we'd chosen the open source version, one of our operations engineers was able to create a plug-in to enable this.

When software didn't do something we needed it to, rather than being forced to change our requirements or choose something different, we were able to code a solution that achieved our objectives.

The Cloud

In the early 2000s, agile development methods rose in popularity. This brought in the ability to break projects into smaller, more manageable sprints. Automated build management followed hot on its heels: continuous deployment felt like an enormous innovation when it was coupled with good automated testing.

At this time, test engineering began to be recognised as a valuable discipline in its own right. Successful online businesses began talking publicly about continuous integration. In 2006 Amazon launched their cloud service, Google and others followed shortly after; the real value of these services were in their APIs allowing them to be accessed by automated systems and infrastructure-as-code was born. Open source; agile development; continuous build and integration; automated testing and infrastructure-as-code were now forming a straight path leading directly to what became DevOps.

At the same time, Facebook was realising that it had an application platform and was actively assisting developers in taking advantage of its data and relationships. In 2008 Apple and Google had both created their app stores and mobile application development was growing at an incredible pace. For the first time since the 1980s, one person could create a complete application or a game, sell it and become a millionaire.

This period saw the largest enterprises re-positioning themselves as cloud service providers. It saw the launch of HP's Helion cloud; EMC's ATMOS storage cloud; and IBM's Smartcloud.

Computer companies, already familiar with the notion that services generated more revenue than boxed products, were jumping on the software-as-a-service (SaaS) bandwagon at a furious pace.

The end of an era

Cloud computing was the innovation that finally rang the death knell for operations as the guardians of infrastructure.

By 2010, anyone with an ability to code at almost any level could launch an online product or service without the requirement for contract negotiations, lease agreements, hardware, or fibre links. Armed with only a credit card and a broadband connection, version control, service hosting, content distribution networks (CDN), application platforms and data warehouses were easily available.

Start-ups were taking full advantage of these new developments. In the midst of a global recession, Facebook and mobile games were a huge growth industry: start-ups needed less capital than at any time previously.

DevOps was already a commonly used term by now, and had even been discussed by government agencies, but there was still no shared consensus of its definition.

Conclusion

When the original computing teams formed, they took on joint roles of both programmer and operator. Their sole focus was the product.

The first commercial computers were entirely proprietary. Programming and operating the machines required specialist knowledge that was rarely transferable except in the most general sense. Organisational structures were created around the computer with little or no reference to products or services.

Programming started to become an open standard in the late 1950s and early 1960s, when Computer Science degrees were first created. For 50 years, Computer Science graduates have left university with only a little practical knowledge of computer operations. Operators have been left to train themselves without the backing of an academic community - or the forums for presenting and debating theories, patterns and methods that developers have enjoyed.

The fragmentation of operations teams into mainframe, midrange and distributed computing teams in the 1980s and 1990s further weakened the operations profession by making it more difficult to share ideas. Now, with the availability of online forums, blogs and Twitter, operations engineers finally have this outlet. As a result, the capability of the profession is growing unlike ever before.

DevOps exists today largely thanks to Twitter and the #devops hashtag.

The availability of standardised hardware and protocols led directly to the creation of the Internet and cloud computing and it's APIs. These all enable developers and operations engineers to, for the first time in 60 years, work seamlessly together to create and operate systems.

CHAPTER THREE

The psychologies of engineers

In this chapter…

We'll examine the impact of limiting teams' roles according to their skill-sets. We'll discover new opportunities for building a healthier more productive business by understanding the product-related tasks undertaken by each team.

To get to this position, we will consider the psychology of the people within the development and operations teams. We'll look at how each team is motivated and discuss some strategies that can lead to a huge increase in productivity.

Using what we've learned about teams within the IT organisation, we'll also examine the essential characteristics of a CTO in a modern online business.

Psychology of Operations Engineers

I've long-considered operations to be something of a vocational career: people find themselves drawn to it and those without a passion for the profession don't last long.

There is no defined career path that leads someone to a role in operations. My own experience saw me take a civil engineering degree, and then go on to a technical call centre before going into testing, then joining the CompuServe UK operations team.

I've met people who have followed the typical development path: with the ink still wet on their computer science degree, they have joined a graduate hothouse and then somehow discovered an operations team and found their place. Sometimes, this happens through a meeting of minds with the in-situ operations engineers, in other situations it's by necessity.

I've also met people who have previously worked as hardware engineers and found that their aptitude for designing solutions and love of software makes operations a good fit. I've seen amazing operations engineers who have no degree, and computer science graduates that couldn't be less suited to the role. I believe the lack of available formal training is one of the reasons the operations discipline is so poorly understood.

So what makes an operations engineer? If you ask one, it's almost certain they'll tell you that they love solving problems. They'll probably regale you with a story about a problem that got them hooked and eager to learn more.

I hate hearing this answer because while it's true, it leads to operations engineers being viewed as people who simply fix system problems at two in the morning. While it is correct that most of us get a buzz from fixing systems, it is merely the tip of the iceberg. As we discussed in the last chapter, operations engineers make an immediate leap from seeing a problem, to considering how to ensure that it never happens again.

Most of the operations engineers I've worked with over the years are also pragmatic; they know no system or process will ever be perfect and they are just looking for solutions that will minimise the number of problems that will impact the business.

This is a really interesting feature of operations engineers' psychology: while they love solving problems, they don't like having to tackle the same problem twice and they crave efficiency.

So, we have a bunch of intelligent, multi-talented engineers who love solving problems and who are happiest when they have no problems to tackle. This leads to a heightened understanding of the operations engineer and indeed, any engineer. While engineers love solving problems, they are happiest when they see their work operating smoothly.

Now consider for a moment what a powerful asset this is: a group of intelligent people motivated to solve problems who are happiest when the business is running optimally.

So why is the operations team often seen as defensive and obstructive? Is it simply a reaction to the conflicting requirements it is asked to deliver?

No: there is one more factor we need to consider.

The clash: operations' need for information

Just as civil engineers must make sense of the ground they're building on, operations teams need to understand the business they operate within.

This means knowing how many people will use the applications they're hosting and when these will be promoted. They need to know the required reliability of a solution; and how efficient and secure it must be.

And these aren't rational desires born out of experience and professionalism; they are almost primal needs. Even so, operations engineers struggle to articulate why they require this information. In fact, they are usually surprised when they are questioned on the matter: to them it's like being asked why they need air.

Operations engineers are frequently handed requirements by product people, or developers, neither of whom necessarily share the same psychology. When an operations engineer starts looking for additional context, their questioning is often seen as aggressive because of poor communication skills.

This can lead to a defensive reaction: the developer or product person assumes the questions are being asked in an attempt to expose an omission. What they fail to realise is, the questioning is a necessity: it gives the operations engineer the context he or she needs to host the application efficiently and pragmatically.

The key here is realising that the operations engineers need more than just the functional requirements that product people generate and give to developers. They need the additional context provided by non-functional requirements (NFRs).

In this day and age, it should go without saying that availability metrics, performance criteria and security standards receive the same attention as features and capabilities. But unfortunately they do not. Too often, operations teams are left building processes to allow them to generate their own non-functional requirements to inform their designs on their own.

Polar opposites: the developer's career path

The career path of a developer is the polar opposite to that of an operations engineer. Almost without exception, developers complete a computer science degree.

Senior IT leaders seek computer science graduates from prestigious universities. Meanwhile, those who attended less well-thought-of schools frequently find their first jobs in hothouse organisations: the major consultancies hire hundreds of graduates every few years.

Clearly, there are exceptions. However, IT organisations have a much easier time recruiting developers than they do operations engineers. There are more developers in the market; and everyone knows what they should expect from graduates given that most have a computer science degree.

The computer science curriculum and the sort of work new graduates are typically exposed to emphasises the functional nature of the applications they create. This is an obvious statement, but it's important: it sets up the first practical difference between developers and operations engineers.

Whereas new operations engineers must understand the context surrounding the applications they host, graduate developers often just need to make sense of the functional requirements. Senior developers and architects take care of the context and ensure requirements presented to the development team take this into account.

That's not to say developers don't care about non-functional requirements. However, it isn't a first or overriding concern as it is with operations engineers. As developers get more experienced, they are increasingly concerned with the performance and robustness of their applications, but their priority will be the function.

Operations engineers often approach this from the other direction, initially requiring a coherent artefact to deploy. What that artefact does is much less interesting than the deployment process and the system's behaviour afterwards.

Very experienced and capable operations engineers understand that the devil is in the detail and learn to care about the function of the applications they are hosting. In this way, skilled and experienced developers and operations engineers share many of the same characteristics - whereas their less capable colleagues have very different attitudes and approaches. This will become crucial later.

Strong leadership: a simple tactic

Strong leadership is key to the IT organisation: it can be the difference between the success and failure of a product. When dealing with senior developers or all but the most junior operations engineers, a simple but effective tactic can be used. Explain to the teams, casually, the problem the business is facing. Don't issue any instructions, just share the problem.

If the organisation is reasonably open about its strategy and current plans, the operations engineers and senior developers will have sufficient context to come up with good solutions.

If it's a simple problem to solve, they will do the work in their own time to create the solution or at the very least, a proof of concept. If it's a larger, more complex problem, they will design the solution and come to you before they begin developing it.

I saw an excellent example of this method in action recently, while I was at Playfish. The time had come to focus on efficient use of infrastructure. I returned from the meeting and was chatting to my team about the shift in priorities, and casually mentioned that we'd be creating some plans for consolidating our infrastructure usage and being more aggressive with capacity management.

The next day, my operations manager came to me with a rough draft plan for some initial actions and some larger initiatives. It was an impressive plan and one that was strong enough to justify bringing in a couple of contract operations engineers.

Together, this little team saved Playfish over $6m in costs over that financial year. My role in this was communicating the changed priorities to the team, emphasising the return on investment (ROI) to the leadership. The team did the rest.

While I was proud of what they achieved, I was actually a little embarrassed by how small a role I played. However, the team was well-informed; had all the context they needed; were comfortable coming forward with developed plans; and committed enough to put it all together in their own time. I had led well.

As we can see from this example, understanding the kind of people in IT can present wonderful opportunities. The engineers can be provided with fulfilling, essential, high-value work. At the same time they will be motivated and part of a working environment that makes them feel successful.

Development: managing distractions

During the working day developers' activities are usually very strictly controlled. Such is the focus on how quickly they can create new features and capabilities that huge volumes of work have been published describing how to manage developers and their work. Numerous project management methods and new roles have been created in order to reduce the distractions they face. There are hundreds of tools available to development organisations all designed to help them manage their work individually and as a team and to develop applications faster with less bugs. I'm trying to create a picture for you of a group of extremely intelligent people whose work is marshalled and whose working lives are actively managed, regulated and monitored. This ensures they can focus on doing what they love and what the business has prioritised. Good development teams also ensure a feedback loop exists in the prioritisation process so that the development team has an opportunity to build the tools and capabilities they need to improve their delivery.

To a great extent, the operations team sets its own priorities. But while there is less focus on the operations team's working day, it is also generally well-managed using a ticket queue.

Typically, internal tickets are generated by developers, testers and other operations engineers requesting fixes or configuration changes. The systems themselves generate tickets, as will external suppliers such as hosting companies, certificate suppliers and payment entities. This is usually enough to keep a couple of engineers busy.

If the organisation is growing, solution and system design and building work will be taking place; that's on top of internally generated projects such as infrastructure improvements, capacity management and security reviews or audits.

However, the tools available to operations are considerably less sophisticated than those available to developers and, as a result, their augmentation and configuration creates a significant volume of work.

So, we have one group of people whose working lives are carefully managed and regulated and another with much more freedom. On one hand is a group of scientists with established patterns and models and on the other, a team of engineers operating with much less guidance and formal training.

From a leadership perspective, the work flows one way: one group is viewed as the creator of the product and the other as a service team. However, the organisation needs both groups to work together as partners if they are to really excel. Here's why:

Infrastructure costs and efficiency

After salaries and pensions, infrastructure costs tend to be the biggest single line item on the corporate budget. The development team creates the applications that need this infrastructure and its decisions at every stage affect the amount and type required.

However, infrastructure is not development's speciality. This sometimes leads developers to choose the most obvious solutions - and these aren't always the best. Earlier in my career while at AOL, I was working on a back-office system which allowed customers to place an order through a web front-end and then spent the next seven to 15 days performing a variety of actions with third parties to fulfil it.

The front-end was horizontally scalable for robustness and there was an application tier connected to a database behind it to store and process the orders. The application tier needed access to a variety of public keys in order to query the partners.

The development team recommended making a storage attached network (SAN) available to the application tier so that the service could access the keys. This is a perfectly valid design, but it would have doubled the cost of the infrastructure and quadrupled the operational complexity: as we were re-purposing existing servers, we would have had to buy the SAN.

As much as I wanted to get my hands on a SAN and improve my enterprise storage experience, I couldn't in good conscience justify the expense, when it would have been a much more pragmatic solution to replicate the keys between the handful of application servers.

It's rare that development makes technically poor recommendations, but very often there are cheaper and more pragmatic alternatives. If operations and development work together as partners, they can achieve the best compromise between capability, cost and time-to-market.

Alerting, monitoring and troubleshooting

Operations teams spend a considerable amount of time concerned with one question: is the application working? Without assistance from development, this is an incredibly difficult question to answer with any degree of certainty: it's very difficult to prove something is working when we are forced to focus on whether it has failed.

Take the simple example of a three-tier web-application, consisting of front-end web servers, a middle-tier application server and some back-end databases. For the sake of this example, we'll assume it's a Java application running on Unix, but the principles are the same regardless of language or platform choice.

As an operations engineer, I can't tell you if the application has definitely failed. I can tell you that the .war file deployed without errors and that Java is running. I can see that ports are open from the web server and that the application server has ports open on the database server.

I can view thread usage on the application server and see that queries are being run on the database server. With a little additional work I can also get statistics on the types of queries being run. I can see the CPU and memory usage, IO wait time, disk use and network traffic in and out of each of the boxes.

But has the application failed? The more of these metrics I collect, analyse and compare, the more confidence I can have that the application has not failed - but I can't be sure.

Now let's consider what we can do when operations partners with the developers. Developers can wrap exception management code around the core application functions. A message can then be sent directly to the monitoring system to say that when executed, the function produced an unexpected result. Now we know, instantly, that the application has failed and we have a particular function to look at.

The benefits don't stop there: there are now monitors with real meaning inside applications. It is now possible for dashboards to make sense to everyone in the business, rather than just the operations team. After all, very few people care about IO wait statistics, but everyone in the business cares about new installs and registrations.

There are many more opportunities than these two but they are enough to demonstrate why the development team and the operations team are a force to be reckoned with when working together and considerably less capable individually.

In reality: uniting development and operations

So how do we get these two very different teams to work together? There are a number of strategies; the efficacy of each is dependent on the number of senior developers and operations engineers in the organisation.

One fool-proof method will work well regardless of how experienced the developers or operations engineers are. For it to work, the operations team must become a development team - at least in part.

Operations teams have been developing applications in secret for years. These come in the form of tools used to manipulate data in log files or find errors as well as those that alert on them, back up and restore scripts. Other applications include automated deployment tools and system build solutions that change configuration based on the underlying hardware. Also applications in their own right are the database queries used to find member records or the tools to generate test data and provide transaction history to customer service.

Operations chooses the languages used to build these applications in the same way as development, considering the availability of those with experience of that language; the tools available for profiling and optimisation; and the libraries that can be used to speed development.

Adopting these methods scores two victories for operations: their applications are better and they earn development and leadership's recognition and respect.

It is, unfortunately true that one can be a developer and know nothing of operations. However, one can't be an operations engineer and know nothing of development.

It is therefore important to implement a formal strategy for application development: follow the team's software development lifecycle and use the same terms. If the organisation demands a formal project initiation document before scoping development work, implement that policy in operations. Adopt the favoured project management method.

Managing operations: compromise and transparency

This will require compromise and finding solutions that work: the operations team isn't going to get project managers, scrum masters and time allocated from the testing team at first. Ensure another operations engineer acts as a tester; pair engineers on projects - ask one person to write the tests while the other works on the application.

And do not be discouraged if this approach appears to destroy progress. It will, at first: initially the team won't see the point in additional process just to change the deployment scripts. However, suddenly that deployment script will work every time. Small inconveniences stop being a factor when a more auditable process is in place.

In addition, ensure all code and configuration files are in source control and accessible to all developers in the organisation. When it's time to create something new or update an existing tool, create a documentation page where requirements are collected; create sub-pages for release notes and links to all the tests. On top of this, encourage everyone in the organisation to review what's being built by the operations team and share its relative priorities, opening up discussion to the whole business.

It will then be obvious that something very dramatic happens when the operations team does business in the open. Suddenly, individual developers start engaging with operations engineers. People drop by who've never engaged with the team before. Corridor conversations start about applications the developers find interesting. Developers offer to collaborate with the operations team. Some even express an interest in working in operations for a short time. This should be encouraged and rewarded.

Developers in operations

When I worked at Playfish, several developers spent time in operations and the experience was rewarding for both sides. Additionally and more importantly for the business, there was a marked productivity boost.

During my time at Playfish, the developer who was taking responsibility for the billing service joined the operations team. By proving herself to operations, she retained her privileged access when she later returned to development. She was then able to deploy her builds herself - which she did faultlessly for the rest of her time at the company.

In cases such as this, developers and operations engineers are earning each other's trust and respect - and the whole business receives a productivity boost. The developers learn what operations needs to know so they can write better tickets; and operations gets an insight into how applications are actually structured, enabling teams to troubleshoot faster and more efficiently.

Dashboards are another great vehicle for promoting the operations team's activities. If information is in a database, creating a professional-looking dashboard with relevant interesting metrics is an easy task.

Any action taken by the operations team can result in a database entry. Using the deployment tools, write time, date, version and deployer data, it's then possible to display details for the last five deployments on the dashboard. This also highlights the teams with buggy applications, as they will require more deployments.

By collecting data from the ticketing system, graphs can be created to show which teams raise the most tickets. On top of this, it's possible to show the number of days since the last security audit; uptime data; and the percentage of successful test restores.

This again shows that the more the operations team does in public, the more everyone in the organisation will engage with the function.

The problem with the CTO's view

Most of the CTOs I've worked with over the last 18 years have a development background. After all, modern organisations employ far more developers than any other IT role.

As previously discussed, development's key activity is creating the features and capabilities that generate revenue. It's therefore no wonder that many CTOs focus almost exclusively on the build, test and deploy process.

On the face of it, this makes perfect sense, particularly in light of the virtues of transforming operations into a development team. However let's take a look at the business from another perspective.

These days most modern IT organisations are building online products and services. These can be public-facing products such as online games, social networks or shops; or back-office business-to-business (B2B) services such as software-as-a-service (SaaS) offerings like Salesforce or Parature.

The applications and services that comprise these offerings are running in the live environment for far longer than they spend in build and testing. Also, while new features, capabilities and bug fixes are important to lift revenue, reduce costs and increase market share, the service currently live is already bringing in revenue.

Is it then appropriate for a CTO to focus the majority of his or her attention on the processes that lead to new opportunities and growth? Surely the live revenue-generating service should receive a significant proportion of the CTO's time?

Yet all too often, information technology is considered static if the product is stable, and efforts then switch to new features, increased market share or new markets. If there are no alerts and revenue is coming in, business operations is considered the realm of the chief operations officer (COO).

But technology is never static. Regardless of how stable the service is and how few amends are made, the technology supporting it and surrounding it is always changing.

"Change is the only constant" is a tired old cliché, but it is a hard fact in IT. If the lion's share of the CTO's focus is on new development, new projects and R&D, he or she is neglecting the part of the business generating the revenue.

In today's world where every service is online, every application has an API for remote access and every company is a worthwhile target for hackers, a CTO must ensure focus is on the current service as well as on new developments and markets. The SDLC is hugely important and consumes vast amounts of resources, but it is being paid for by revenue generated by the live products.

Therefore CTOs need to ensure that new developments and initiatives are built with the hooks necessary to report on current performance. Transactions should be visible in near real-time on dashboards, not just on reports processed overnight by data warehouse systems. Infrastructure costs shouldn't be assessed merely when new budget submissions are required but should be managed each day. Daily operations and support functions shouldn't just receive attention when problems occur.

Conclusion

By understanding the mindset and the context within which operations and software engineers work best we can see how well they can work together if we remove the barriers to cooperation. These barriers are the artificial team structures we put these people in. They are the inappropriate focus on building and launching these products rather than the entire product life-cycle including management, support and decommissioning of the products.

By focusing on this product lifecycle in it's entirety IT can make itself understood by the writers and implementers of company policy. Rather than being faceless software engineers, operations engineers and test engineers deployed as needed, we become the people who build, launch, manage and support each product. We become associated with the product; we become the owners of the product. We take responsibility for the product's failures and consequently it's successes.

This creates it's own motivating feedback loop. Product failures cause everyone in the team to feel responsible and double down their efforts to succeed. Today this only happens in start-ups because it's only there that engineers have any significant voice in the product's direction. Product successes motivate the team to even greater achievements because the success of a product is a direct measurement of the team's success.

In the last ten years it's became possible to code infrastructure and tests in the same way we've been coding our applications for years.

This new development gives us the last nail we needed in the coffin of the old IT organisational structure. It frees us from any reason for segregating software, operations and test engineers and allows us to focus all of our engineers on the products we want them to create.

CHAPTER FOUR

Management theories and how they relate to modern IT organisations

In this chapter...

We'll examine age-old management theories and show how they are still relevant to the modern workplace. Beyond that, we'll demonstrate how to encourage real, focused productivity from engineers, while solving the motivation and job satisfaction problems that plague medium and large businesses.

Finally, we'll take a look at how company policy-makers are failing the IT organisation and consider how company policies need to change.

Line management: a near-impossible task

Most successful organisations can perform day-to-day business activities with focus and dedication and as a result, achieve reasonable productivity. Equally, most companies can create enough space to form coherent strategies and day-to-day plans.

But many businesses struggle to consistently translate long-term strategies into medium-term action plans that put employees' day-to-day work into context.

More often than not, managers are ex-engineers who have received little or no training. This is in spite of the fact that line management is the hardest job in the technology department. After all, these are the people tasked with turning dreams into reality.

This lack of training and preparation causes problems for both the engineers being managed and the department heads and leaders. It's impossible for leadership to create tactical implementation plans that achieve strategic business objectives if they spend a significant amount of time dealing with employee motivation, communication and organisation.

To make matters worse, it's very difficult for team managers to provide context for employees' work and communicate how it contributes to the business strategy if the leadership isn't providing tactical plans. So both problems exacerbate each other.

Two-factor theory

Published in 1959, Two-Factor Theory - or Herzberg's Motivation-Hygiene Theory - states that certain conditions motivate staff, while others can have the opposite effect. The man behind it, US psychologist Frederick Herzberg, published his theory after conducting hundreds of interviews with engineers and accountants, building on the work of Abraham Maslow. In recent years, Herzberg's theory been regarded as a simplistic model, but it is this lack of complexity that makes it so useful to management.

Herzberg's research led him to conclude that the factors motivating employees usually stem from the work itself, or effects arising from this. In contrast, he found that causes of dissatisfaction stem from the work context.

He found that achievement, recognition, the work itself, and responsibility and advancement were factors that cause the greatest sense of satisfaction or motivation in employees. The causes of dissatisfaction that de-motivated employees included company policy and administration, relationships with managers and business leadership.

Interestingly, salary was considered as both a motivating factor and a cause for dissatisfaction. Herzberg commented that when salary was listed as a de-motivating factor, employees were concerned with its administration as well as the company compensation policy. When salary was listed as a cause of satisfaction, it was frequently tied into recognition and achievement. Herzberg and many people who interpret his work consider that the money each employee is earning only boosts motivation in the very short term at best. In the long term, if someone is paid appropriately, the amount they take home is irrelevant to their job satisfaction or motivation.

So according to Herzberg, developers, operations and test engineers need to be paid appropriately. They are motivated by achievement, recognition, the joy of their work and having appropriate responsibility levels. In turn they are dissatisfied by company policy and administration, and the relationship they have with their management and leadership.

Does that sound familiar?

To make the most of this information, the manager needs to be free to recognise achievements; facilitate a forum for peer recognition; tailor the work that each employee undertakes; and provide a career path for engineers. At the same time, managers need to be able to relieve the stress of company policy and administration from each engineer, while developing strong relationships with both staff and senior leadership teams.

So why do so many companies provide only one opportunity for recognition and advancement per year and sanitise that process to the point where everyone feels a sense of dissatisfaction throughout? Large organisations often lose focus on the end goal, their reason for existing in the first place. But for some employees, the end of year review is the focus. This process becomes more important than the product the company makes, or the revenue it generates.

In large organisations, different groups of people have contrasting end goals. Skill set silos encourage this behaviour: from marketing through to HR, each team wants to excel. But they don't build the product, which makes them focus specifically on their own departments. To software, operations and test engineers, this is a huge distraction. They are surrounded by groups that impose seemingly irrelevant products and policies on them. This is fertile breeding ground for dissatisfaction and de-motivation.

Maslow's hierarchy of needs

In 1954, Abraham Maslow published his theory of the Hierarchy of Needs, which stated that people are motivated to achieve a state called 'self-actualisation'. Maslow considered that first, people need to secure some basic needs: air to breathe; water to drink; food to eat; and shelter.

Having achieved these goals, they would then be motivated by personal safety and health and after this, friendship and intimacy, followed by self-esteem and recognition. Finally, the individual would then be motivated to reach their full potential. While one person could pursue physical fitness, another might choose to excel in their profession.

Maslow became fascinated with the successful people around him, starting with his mentors and famous people of the day, such as Albert Einstein and Eleanor Roosevelt. During this time, Maslow wrote much on the subject of self-actualisation. Included in his definitions are words such as 'honesty', 'playfulness', 'organisation', 'completion', 'rightness' and 'effortlessness' - qualities we generally admire in successful people.

For someone to be capable of working in a modern business, they will have already achieved Maslow's first three hierarchies of needs. This presents an opportunity to empower these people to discover their true potential and apply it. But in order for that to happen, the organisation must provide the environment, work and structure that encourages staff to gain self-esteem and receive recognition from peers and leaders.

So why do we obsess over one-size-fits-all performance reviews? Why are award ceremonies and opportunities for peer recognition so gut-wrenchingly embarrassing, with carefully-worded announcements and forced applause? Faced with the difficult challenge of recognising employee achievements, organisations seize upon the first idea that at least offers a response to the problem, even if it doesn't offer a solution.

We have created a culture both in our education systems and at work that sees achievement measured by the completion of goals, such as passing exams or the end of a project. This sees firms recognising employees' individual achievements through a single moment of recognition, such as a little award, a gift voucher or a certificate.

At best, that employee has been given a short-lived boost. At worst, they've been presented with another sign that the organisation has no idea how to manage them, breeding more dissatisfaction.

Theory X Theory Y

In 1960, Douglas McGregor published 'The Human Side of Enterprise', proposing 'Theory X' and 'Theory Y'.

According to Theory X, management assumes that employees are intrinsically lazy, work solely for money, and dislike what they do. Management's role, therefore, is to control employees, ensuring they deliver by using coercion and financial incentives. Theory X says that managers can only oversee a small number of people, as they are also effectively quality-controlling the work of the entire team.

Theory Y assumes that staff are intelligent, self-motivated, enjoy their work and take on the organisation's objectives as their own. According to Theory Y, managers can lead a larger group of multi-disciplined people, providing guidance to the team and giving one-to-one coaching as needed.

The people found in IT departments fit almost without exception into Theory Y. That is until they are forced into Theory X by company policies, administrative burdens and the difficulties they face trying to overcome the silos within their organisations.

Spheres of control and influence

'Spheres of control and influence' is a concept that is introduced to managers looking to transition into leadership. It is most commonly taught during a structured question and answer session where students are encouraged to concentrate on those elements that are within their absolute control and list them separately from those that they only have influence over. Inevitably, the student realises that the only control anyone ever has is over themselves and everything else is just subject to their influence.

The next exercise in the curriculum is usually around influencing skills. This session tries to equip the students with tools to help navigate projects through the organisation.

In the traditionally structured IT organisation, these exercises are incredibly valuable. The software, operations and test engineering teams all have different day-to-day agendas and while, in theory, their medium and long-term goals are the same, they often aren't.

IT leadership isn't necessarily best placed to help line managers push initiatives through. Therefore, any tools that can help complete the work are obviously a benefit to the organisation as a whole.

But is it right that organisations don't lend themselves to coordinated productivity? Is it now assumed that the normal state of teams in our organisation is to work against each other? Are we forced therefore to equip them with influencing and negotiating skills in order that they can achieve the business' objectives?

This is a further evolution of the idea that process and policy has greater importance and is of more immediate concern than the company's goals. When an organisation equips its people with skills to help them work around the obstacles the organisation has itself placed in their way staff receive a clear message of confusion and disorganisation.

S.M.A.R.T.

In 1961, President John F Kennedy delivered a speech before a joint session of congress announcing: "I believe that this nation should commit itself to achieving the goal, before this decade is out, of landing a man on the moon and returning him safely to the Earth."

In one sentence, President Kennedy stated the goal of one of the largest, most expensive, ambitious, technically complicated and dangerous projects ever undertaken. If that's possible, why do we in IT find it so hard to specify and invest in our goals?

At every organisation I've worked for, HR has mandated that goals are set regularly and employees' performance is judged on reaching these. I've developed numerous strategies for dealing with this over the years. At first, I worked very hard to convince my boss that I had achieved the goals he'd set, even when I patently hadn't. It was a great exercise in creative writing.

As I got better at handling performance reviews, I evolved another strategy. I wrote about how the goals I'd been set had been made irrelevant by market changes and changing company strategy. I then detailed what I had actually achieved that year.

When I became a manager, I learnt that I had the freedom to change the goals as necessary and so my team and I would set our objectives three months before year-end. Of course, we always had a phenomenally successful year, regardless of department or company performance.

As I became more senior, leadership and senior HR representatives tried to convince me to get on board with the process. They would explain that, of course, it's a bad system but it's necessary: how else would managers be forced to confront performance problems and achievements be recognised and rewarded?

Yet the tools available for setting goals are excellent: SMART - which stands for 'Specific, Measurable, Achievable, Realistic and Time-based' - is phenomenally powerful. If we examine President Kennedy's Apollo programme statement, we can see that it has all these qualities. The man hit all five objectives in one sentence - and he was talking about a project which was orders of magnitude more complicated than any faced by IT departments.

If objectives are culturally essential, institutionally ingrained and we have great tools and inspiring examples to follow, why have I spent my entire career dissembling about my goals? By now the answer will be obvious: these goals were ancillary to my work; they were company policy and procedure. They weren't necessary to get the job done or to achieve the recognition from the people I respected - or even to achieve financial rewards. Goals and objectives were another distraction from what I enjoyed and was motivated to excel at.

Group Communication

It's common for medium and large organisations to perform employee satisfaction or health surveys: all the organisations I've worked for took these very seriously and there is huge pressure from leadership to achieve a high rate of submission. Once the results have been analysed, the pressure builds on leadership to interpret the results and put action plans in place.

Every time these survey results are reviewed, there is one constant: employees want more communication. Every time this result is presented, leadership seizes upon the easiest and least complex solution: more awards, presentations and Q&As - all of which are irrelevant and fail to actually address the problem.

Unfortunately, surveys always present the question as: "Rate how well management communicates company objectives on a scale of 1 to 5, where 1 is very dissatisfied and 5 is completely satisfied." This question is never one with a free text field.

Leadership fails to realise that employees' needs are simple: they want clarity on how their work directly contributes to company strategy.

In other words, staff want to understand the context within which they operate. They want to know that what they do matters; they are looking for another opportunity for recognition and self-actualisation. But all-too-often, they are instead given more company policy, process and boredom.

Limiting the dissatisfaction

Developers, testers, desktop support and operations engineers share many of the same motivating factors. Generally speaking, all are driven by the respect of their peers and learning new technologies. They are also motivated by a feeling of contribution, whether to the business or their profession as whole.

But once a year, each year, all these motivating factors are overshadowed by money. It has long been understood that money is not, by itself, a long-term motivator, but it needs to be sufficient that employees aren't de-motivated. However, getting salary levels right is very difficult. Particularly because it isn't the salaries that are important. Starved of any other form of recognition salaries are the only way employees can judge how valuable their contribution has been.

However it is important that people are paid enough that salary doesn't become a source of dissatisfaction. Most IT managers have had an argument with HR over the salaries of their people. This is often prompted by the fact that, generally speaking, HR under-values the salaries of IT staff.

Examining the process used by HR for setting salary levels reveals the cause of this problem. Most companies define a percentile range for salaries. If the business is investing and looking to increase market share, it will instruct HR to aim nearer the top of the salary range. Equally, if the firm is aiming to reap the rewards of its investments and reduce costs, salary levels are set further down the range.

HR then performs market rate research, pitching the salary band for each role according to the guidance received from leadership. HR will consider the roles using the job titles that have been assigned to them. This is one of the reasons why HR is so keen to align job titles with industry standards - and it also makes filtering CVs easier for recruitment.

If the organisation employs Java developers and calls them 'software engineers', HR reviews employment listings in similar industries for the same job titles that list Java as the primary language requirement. This reveals the first problem: the industry searched for might not be relevant to the particular organisation.

When I was managing the UK operations team in AOL about 10 years ago, the bulk of the IT organisation was developing, testing and maintaining the Java-based broadband ordering and provisioning system. I was struggling to hire new engineers and my team rightly felt that their salaries were too low.

I took the problem to HR, and in spite of being aware of my recruitment issues, it performed market rate research analysis and concluded my team was paid adequately. HR had looked up system administration roles in telecommunications companies, but the roles in question were Unix system administrators supporting network inspection tools. My team ran a heterogeneous estate consisting of Sun, Linux, HP-UX and Irix, supporting web/cgi and Java applications, as well as Sybase databases. Yet I couldn't convince HR that the roles they found were irrelevant.

I did my own research, collecting job descriptions that more accurately reflected the eclectic roles of my team and successfully demonstrated that a market rate adjustment was needed. This was time-consuming as these sort of multi-skilled operations roles weren't common at the time - and it was only necessary because HR didn't understand the nuances of the roles in the IT organisation.

Not all Java development jobs require the same level of knowledge, skills and experience. It's one thing to build low-volume web applications in Java and quite another to build optimised high throughput transactional services. This nuance is extremely important in setting the salary levels for this sort of role.

This highlights a need for HR to take its role seriously, and ensure it understands the detailed responsibilities of individuals, as well as the personality types and motivating factors of the people in those roles.

Peer recognition

Assuming salary is now, for the most part, a solved problem, it's necessary to look at more effective ways to motivate people in the IT organisation. These methods should be supported by the business because they don't cost any money, yet implementing them can meet with incredible resistance.

Peer recognition and respect is a powerful force, but most organisations do little to encourage it. Those that do make an effort usually end up with a cheesy, company value based system which encourages nominations and has an award ceremony.

But peer recognition doesn't have to be part of a programme containing posters and pictures of teams grinning manically. It can be a genuine, inclusive experience that encourages everyone to participate without the threat of 'mandatory fun'.

I've seen teams successfully offer lightning talks to allow people to share the new things they've been working on. This effort would have been much more effective if it was sponsored by leadership, which could provide context for the talks: having a senior IT executive introduce and say a few words about how the discovery or invention contributed to the business makes a huge impact on everyone's morale.

Even those attending the talk are motivated by the knowledge that one of their peers is making a difference and the chance is there for them to do the same. This then creates another opportunity: rather than holding a popularity contest to create peer recognition, leadership can nominate the individuals or teams whose inventions or discoveries contributed to the success of the business. This not only motivates people, it empowers them by providing the context for the contribution. Employees now understand exactly what moves the needle and will naturally start thinking about how they can help.

The open source opportunity

Peer recognition shouldn't just be confined to fellow employees. Most firms are using open source software, but very few contribute anything back. Equally, most medium and large businesses are claiming R&D tax relief but contributing little, other than to shareholders.

Imagine how much time has been spent in every IT organisation in the world creating log processing scripts, dashboards, backup solutions, build systems and artefact management solutions. Every IT firm in the world needs some or all of these and they are usually built in-house because there isn't the budget to buy a commercial-off-the-shelf (COTS) solution.

Imagine if these companies had contributed their efforts back to the community: motivation naturally increases if staff are able contribute something to their profession. Imagine further how IT would feel to be invited to speak about its contribution at industry events and the recruitment opportunities this would open up.

Now, visualise how much time will be saved by every company when open source software for all of these ancillary services is available and simply needs tweaking, rather than creating from scratch. Contributing to open source isn't an altruistic activity because the opportunity exists for that project to receive contributions from other companies and private individuals.

This means one company doesn't have to bear the brunt of bug fixing and feature development costs: the expense is shared among all contributors. Ultimately, the company that creates a new open source project gets the benefit of many more developers in addition to its own staff.

There's no reason why development that isn't directly related to the core product shouldn't be open source. There's no revenue or IP loss to consider and only cost reduction to be gained.

Recognising IT's contribution

Recognising IT's contribution has a motivating impact and should therefore be utilised throughout the business. Yet many IT organisations are doing a poor job of communicating with their departments.

When quarterly financial results are announced, huge press events take place as PR and marketing kick off frenetic preparations. Products that have spent a year or more in secret development are launched to great fanfare and a flurry of emails about revenue, market share and share price are sent.

But all the while, IT staff are given no explanation of how they contributed to the results. Instead, there's usually a deafening silence regarding the impact the announcements will have on their work and headcount.

This is also true when product launches are hideous flops plagued with technical problems. The silence remains in spite of market downturns and global recessions, as post-mortems and post-implementation reviews are conducted.

Often, staff use health surveys as an opportunity to request more communication from their superiors. But nothing changes: the leader of the department simply reads out the emails, asking if anyone has any questions.

The role of those derisively called 'middle management' is, in part, to turn high-level company strategy into implementation plans. They should be able to provide the context for their teams enabling them to deliver applications, services, features and capabilities that can be marketed and sold.

Yet when it comes to interpreting company, industry and global events and putting them into context, only PR-sanitised emails from the CEO prevail.

For every company leader complaining about a lack of commitment from employees, there is a leadership team failing to distil strategy into understandable implementation plans. How can there be commitment when the IT organisation doesn't understand how its work is relevant to the business?

I saw a good example of this during the dial-up internet access days, when I was working for AOL. British Telecom (BT) introduced a scheme allowing internet service providers (ISPs) to purchase lines in bulk. It called for a new system, and less than a dozen of us were assigned to working on it.

It was a very difficult launch, calling for five builds before it was stable and free from problems. It saw us all working very long hours under intense pressure and scrutiny.

A few weeks later, I was in the COO's office during a meeting and noticed a printed email on her desk regarding the project. I asked if I could read it: the email stated how much money had been saved by the new system.

That small system built by half a dozen developers, testers and operations engineers had saved, in costs, around a quarter of the total revenue for the year.

I asked if I could share this with the team. Although the COO had no problem with it, she hadn't thought it was necessary: she had not considered the motivational impact that kind of information would have on the IT organisation.

Needless to say the developers, testers and operations engineers were blown away by the impact of their contribution. And, ultimately, it led directly to a better working relationship between us as we basked in the glow of our success.

Conclusion

These tried and tested management theories give very clear guidance on unlocking teams' potential. The sources of de-motivation and dissatisfaction are there for all to see - and they are almost universally acknowledged and accepted.

There are great tools for setting achievable objectives and providing mechanisms for recognition, yet the whole process has been turned into a burden. Employees are crying out to use their potential for the organisation's ultimate gain, but their views are currently just another box-ticking exercise.

And so, by recognising that people are motivated by the work they do and peer recognition, acknowledging that company policy and administration is actively working against this, how can firms free themselves from this rut?

To lay the foundations, teams must be aligned so that they can never work against each other. Their working environment must reflect this, smoothing the way for them to achieve their potential and achieve business goals.

This can be done using Next Gen DevOps: the idea that software, operations and test engineers, as well as product managers, marketing and sales, become product-focused.

Teams should create their own objectives, including within these revenue, profit and market share metrics. Objectives can then be broken down into responsibilities for each individual, with this tied directly to business success. This ensures each employee sees how their contribution affects their team's.

There is no point focusing on one-off, fire-and-forget goals graded once a year when there is a far better opportunity. Almost every other day, each individual team member will be achieving something specific, measurable, achievable, realistic and time-based. There's no need for additional forms, procedures and systems.

It's time to give multi-discipline teams the opportunity to talk to peers within and outside the organisation about how they work. After all, who would turn down marketing and recruitment for free? Meanwhile, each employee will grow and reach their full potential, channelling all of that capability back into the business.

CHAPTER FIVE

The role of bribery in IT

(Or why people need to negotiate with each other to do what they're all paid to do anyway!)

In this chapter...

We will look at how the conflicting priorities present in the IT organisation can be aligned. This chapter will demonstrate that software, test and operations engineers all have something to offer each other and they can work synchronously, delivering better quality products, faster.

We'll also find an opportunity to further improve productivity, quality and commitment gains just by looking at priorities in a slightly different way.

Conflicts within the IT organisation

Looking again at the conflicting priorities within the IT organisation, it can be seen that:

• IT leaders are looking for predictability and constant movement. They want high quality software releases built and deployed rapidly. This sees them pushing the boundaries as they strive for reliable, cost-effective products.

• Developers want to build fast, robust applications quickly. Once one story is closed, they want to get straight on with the next one. They require that code is submitted instantly, with builds delivered and tested as quickly as possible. They want to make sound design decisions so they don't have to wade through technical debt in the future.

• Test engineers want to build an exhaustive suite of innovative tests that deliver the quality levels required, but do not slow down the build and release process.

• Operations engineers want high quality, reliable services, with infrastructure running efficiently. They aim to strike the ideal balance between flexibility, reliability and security.

So, looking at these conflicting priorities, how can rapid movement ever be predictable? Is it possible for developers to make fast progress through their tasks and not fall foul of strategic mistakes? How can the business thoroughly test applications, find bugs and avoid slowing down the development of new features? What is the best way to balance flexible use of infrastructure with efficiency and security?

Overcoming conflicts

Typically, leadership looks to project management structures to align these conflicts. But this leads to short term compromises that amalgamate in much larger problems in the future. It's for this reason that we see organisations failing to refresh their infrastructure when it reaches end-of-life and internet accessible servers not patched and getting outdated.

In the worst case scenario, development and testing teams are played against each other in a battle over how much testing is enough. Developers who fall foul of this find themselves arguing about whether the unit tests already in place provide sufficient coverage, all the while knowing these tests were never designed to provide this level of assurance.

Test engineers find themselves squeezed between immovable launch dates and over-running development tasks and are forced to compromise, knowing that they will take the hit for any bugs that make their way into the live environment. The development, testing and operations teams find themselves at odds over infrastructure function, performance, application design, capacity, load testing and even how applications will be deployed.

In a desperate bid to align their goals with those of their peers, this sees frustrated team managers ask for project prioritisation. They know that this isn't the ultimate solution, but hope they can negotiate with their colleagues in other teams to minimise the conflict. And how many times have high priority projects in one division of an organisation been effectively cancelled because of the lack of cooperation from another division?

At the same time, insightful IT leaders look to modern automation techniques in an attempt speed up the performance of their teams. The theory behind this being: if operations can provide infrastructure as needed, builds will be executed more rapidly. Software engineers will then commit in smaller increments; each build will have fewer new features to test and hence the cycle of bug identification, fix and re-test will be quicker. They hope that reducing the time spent on infrastructure provision will free up operations to assist development: many people think of this as DevOps.

Although in practice, this approach can yield results, it is at an ever-increasing cost. Infrastructure is of course a regular monthly expense - and removing it is far more difficult and time-consuming than adding it. Additional environments increase spend; plus they need more management and maintenance. They create configuration management challenges that drain momentum from the operations team and distract them from their non-project specific tasks, such as security and data management.

The Sims Social team at Playfish were under tremendous pressure to keep up the momentum of adding new features and content to the game. The game team grew and grew until they were working on several major features and content updates at once. This meant that the team needed three times as much development and test infrastructure as other games teams. On it's own this wouldn't have been a big problem as the games team's didn't normally require that much infrastructure.

Then problems were found in one particular build. The scrum master for that release decided they needed more infrastructure so development could continue uninterrupted while this particular problem was investigated. In the next sprint a new feature was proposed that would interact with back-end services games didn't normally interact with. This required a full system integration environment with all the back-end services represented. A few more sprints came and went and upon review it was found that The Sims Social development and test environment was now 73 Amazon instances.

This was a time when Playfish was trying to be efficient with it's use of infrastructure. Reviewing the use of the environments with the team, determining which components weren't necessary and planning when environments would no longer needed took weeks. During this time the team got used to using all these different environments. Then the question arose as to whether the team actually needed all the environments now. Did convenience equal need? How much time was being saved because of the new environments? Did the savings realised from the saved time outweigh the monthly cost of the infrastructure? These questions grew and grew and the operations team was dragged into providing cost forecasts and confirming whether these instances would be counted against the reserved pool.

A couple of problems, solved by a project management structure working to tight deadlines ended up dragging engineers, scrum masters, managers, directors, executives and the finance department into weeks of questions, calculations and reviews. Meanwhile infrastructure costs added up month after month. By the time a decision was finally arrived at and appropriate infrastructure designed for the team three months had elapsed but far more importantly than three months of infrastructure cost was the disruption caused to the game team, the operations team and the leadership group.

Lessons learned from a Start-up

IT's role is to create products and systems that can be used by other teams to realise revenue. This is true whether the business is a social gaming start-up, the largest insurer in the world, a national utility, or a manufacturer. The software or service created by the IT department is often the revenue-generating product itself. If not, it's likely that the IT systems enable greater capability, productivity and quality, increasing revenue opportunities and reducing costs. It therefore seems obvious that the product should be the primary focus of the IT department.

And at successful start-ups, it is. A group of people with different skills, experiences and backgrounds are brought together to create a product that disrupts the market or creates a new opportunity. That team focuses entirely on the product it is creating.

Executives in many large organisations express admiration for the innovation and performance exhibited by start-ups. More than once, I've heard senior IT leaders express a desire for their billion dollar organisations to behave more like start-ups. It's reasonably common for large firms to spawn smaller, semi-independent companies when they need to try something new. This is what I found at British Gas' Connected Homes. Realising that there was a new market opportunity, the utility wanted to create a business that used a start-up approach.

Overcoming silos

In spite of the opportunity this start-up mentality creates, parent organisations continue to silo their IT organisations by skill set. In contrast to start-ups, these siloed teams have no way to contribute directly to revenue or market share. They are incentivised differently from one another, further exacerbating conflicts. This leaves them feeling isolated and helpless and over time, it creates the siege mentality we've all seen IT teams in larger organisations' exhibit.

The only way these teams can make progress is to overcome organisational structures, building relationships with their peers. However, as soon as priorities or managers change, or products fall out of favour, all this good work is undone. The cycle begins again when the new team, with conflicting relationships and different tools, starts work on the next product.

Even high-performing start-ups fall foul of this paradigm eventually: success will lead to hiring more people. Almost inevitably, it leads them to ignore the structure and focus that made them successful in the first place and instead adopt the traditional organisational paradigm.

Almost immediately, the only progress available is incremental, time-consuming and fraught with organisational problems. These are categorised as cultural issues, kicking off years of discussions about retaining the start-up mentality, that result in some perfunctory, but largely ineffective actions.

Lessons learned from the billion dollar business

The previous decade saw a focus on developing capable software systems. The software development lifecycle (SDLC) received a huge amount of attention; seeing various frameworks created to help IT organisations find their path to repeatable success. This has seen the formalisation of Rolls Royce's Waterfall method, as well as Agile Development and IBM's Rational Unified Process (RUP) among others.

Moving into the present day, social, cultural and technological changes have now turned software systems into online services. In 2014, there's almost no sign of stand-alone software: even Microsoft's Office is a cloud product.

All the same, these SDLC frameworks are useful in helping organisations build software. But after that, firms are on their own: the frameworks provide no guidance for the period after the product is launched. The Broadband Management System we built at AOL ran for eight years. The successful games we built at Playfish ran for more than three years. Facebook has been live for the general public since 2004. Amazon EC2 has been live, after a fashion since 2006. These successful online services live and generate revenue for years after launch.

These successful services launched with a minimal feature set. They were then subject to constant new development and bug fixing throughout most of their revenue-generating lives.

The most successful services live long enough that significant technological changes occur, necessitating further amendments. After all, Amazon EC2 is using significantly more capable infrastructure now than it was in 2006.

Every product is conceived, designed, built, tested, implemented, launched, managed, supported and eventually decommissioned. SDLCs can help get a product ready for implementation in the live environment but they don't launch, manage, support and decommission it. This gap is filled by the product lifecycle.

Conway's Law

In a paper entitled "How do committees invent" published in April 1968 a programmer called Melvin Conway stated:

"organizations which design systems ... are constrained to produce designs which are copies of the communication structures of these organizations"

This came to be known as Conway's law.

Now consider that when Conway was writing software the systems he and the team he worked in would create software to run on a single machine. A programmer in those days would have just their code, a compiler and the system it would run on to contend with. Contrast that with the work of a modern software engineer who has their code and version control systems, build systems, an array of third-party sourced libraries and a variety of other complete external systems to interact with over a number of potentially different network systems and millions of users with completely different systems and operating systems.

That certainly lends weight to the idea that if we require software engineers to create products that need to interact with other products then perhaps we should create product teams.

As successful start-ups can testify, making products the key focus can reap huge rewards.

Knowing this, organisations can get ahead by embracing the fact that in today's modern world, products are online services with their own lifecycles.

By adding the operations products approach proposed in Chapter 1 into the mix, it's possible to create some guiding principles:

• The products are the organisation's focus: any structures created must encourage this.

• Every product today is an online service that must be built, launched, managed, supported and one day decommissioned. In other words, every product follows a lifecycle.

• Tools, services and scripts are products in their own right.

So, when building new IT organisations, or assessing existing ones, it's integral to start by considering the products, as well as the skills and capabilities needed to build, launch, manage and support them.

Rather than forcing development, testing and operations to negotiate with each other, create product teams. This can be done by thinking of software, operations and test engineering as skills or capabilities.

These teams can be populated with the skills needed, rather than with specific roles. In some cases, the product will create a need for specialists but in other situations, there will be an opportunity to look for generalists with a wider range of skills.

By doing this, there is no need to disrupt other teams to pull in skills in case of a critical incident or major launch. There's no requirement for these teams to negotiate with each other or have their progress blocked. Instead, they can transition seamlessly from build to launch, into management and support and finally when the product has served its function, it can be decommissioned.

The product teams

The skilled teams will now grow to understand that the products they build, launch, manage and support are their responsibility - and they will constantly monitor their performance.

When the products are the tools the organisation needs for its projects, the users are the heart of the solution. If another area of the business needs a product, the discussion it has is with a team that also uses it, rather than a group that merely provides it as a service.

Revenue generation

Most IT organisations have to build, choose, manage and support a huge matrix of products. Some of these generate revenue directly;

while some support the products that make the money. Both should have equal priority, but they often don't.

Development environments are an excellent example of a product that supports revenue generating products. Without them there would be no product and no revenue. These are systems, similar to the live environments, but there are differences: they might be less powerful; they are probably hidden behind networks or firewall rules. So how do these get built? How are they supported? Who manages and maintains them? And who is responsible for ensuring they work as they should?

For many organisations, these simple questions have very complicated answers. Some operations teams won't provide or support development environments because they require much too more work. This responsibility is passed on to the development teams. But the development environments show little regard to the live environments; they are often built by teams with no detailed knowledge of this area. Who would know if one of these environments was compromised? They are not always monitored and only accessed by applications deploying builds onto them.

This happens because for many organisations, it is only the spending that occurs on directly revenue-generating products that is worth the investment. Anything else is merely a cost and must be limited.

But this is an incredibly short-sighted view. It comes from the assumption that IT is a service to be rendered to the business, when in fact, it's just another function. Spending on products and services that enable greater revenue opportunities is just as much an investment as spending on those that directly generate cash. Therefore, all products require appropriate investment.

The product team approach can also be applied to the provision of good quality development environments. This sees the team that uses the products also manage and maintain the environments. After all, by using this approach, each product team should have an abundance of the skills required. Further, each team can be incentivised to do this efficiently.

But who is responsible for ensuring they work as they should? Again, the product team can take this on. They have the skills they need; and there are tools available which are themselves owned by a team of people with the necessary expertise to build, launch, manage and support them.

Supporting Development Environments

Development environments can be thought of as supported by the application of a number of products simultaneously:

- Configuration management
- Continuous integration
- Monitoring and dashboards
- System integration
- Data management
- Ticketing and documentation

If each of these is owned by a team of people with the appropriate skills, products can be built on environments that accurately reflect the live state. Each member of the product team knows the live environment intimately and is responsible for defining, or at the very least reviewing and approving its configuration. Likewise, the product team is responsible for defining, reviewing and approving configuration of the development environment.

By introducing product teams into this mix, conflict becomes a thing of the past. This sees an end to the argument that the development environment isn't exactly like the live environment. No longer will operations conflict with development teams over the behaviour of environments. The product team knows its environments and the differences between them - and it understands the implications of these.

Diversifying Product Teams

Any software project will require as a minimum:

- Integrated development environments (IDE)
- Source control
- Build tools
- Application repositories
- Hosting infrastructure
- Behaviour driven development tools
- Configuration management systems
- Orchestration mechanisms
- Data retention and management tools
- Documentation systems
- Ticketing systems
- Task management tools

And that's before considering the different languages employed; database technologies; third party-provided libraries; and any legacy that needs to be supported. So it's really not possible to build a team for each one of these products.

Instead, firms should look for commonalities. This could mean grouping tools together by skill set. It could be beneficial, for example, to create a team to own all the products that require a particular language such as Java or Ruby.

Organisations could create a team that owns all the products in use by particular stakeholders. Companies could group together all of the products required by the marketing function to be owned by a single team. Or, firms might choose to group together the products required to develop new ones. Any further development of these tools would then be owned by this team.

There are many options, but what's important is defining a strategy for product ownership that team managers can use for guidance - and make it one that eliminates conflict.

Increased Productivity

This strategy sees the power to choose where the IT organisation's resources are deployed firmly back in the hands of its leaders: compromises aren't made without their knowledge.

Rather than sidelined service teams making tactical priority calls in the moment to satisfy an unsympathetic project structure, enfranchised product-focused people with clearly defined responsibilities and appropriate project and management structures can make decisions with help from well-informed leadership. These decisions will be informed by a matrix of well-understood product relationships.

Individual teams will use each of the products and services in subtly different ways. This will see them discuss the products and their use with each other, sharing new discoveries and techniques. Deficiencies will be discussed in every corridor and petitions will be made to the project and management structures, calling on them to allocate time to resolving these. Bug fixes and new developments will be celebrated by everyone in the IT organisation.

As Chapter 1 showed, in most organisations, the operations team is forced to compromise essential services such as security and data management. This is necessary to ensure they are responding to the needs of the software engineering teams as quickly as everyone wants them to.

But that whole situation can now be turned on its head. Security and data management are now provided by the solutions and services owned by product teams. These are comprised of software, operations and test engineers as required and their time allocation is managed by the project management structure. Now any compromises are made above the counter and in the full knowledge of everyone in the organisation.

This strategy of coordinated priorities is one method in which product teams experience increased productivity. Deployment delays caused by other products experiencing high severity incidents become a thing of the past. There will be no testing delays because one project is in crisis and needs all hands on deck. There are no conflicts over the quantity and function of hosting environments, or the provision of any other products and services that the team needs.

The delicate art of cat herding

As focus moves away from the software development lifecycle towards the product lifecycle, IT leaders are presented with new opportunities.

When operating in siloed IT organisational structures, the leadership is forced to step in to resolve conflicts, remove blockages and change priorities almost daily.

If the standards of the IT organisation are defined by the configuration of the products it uses, transforming the way the business works is a matter of changing these configurations.

For example, infrastructure implementation is a matter for the configuration management product. Those looking for cost reductions can choose to make changes to the configuration management product itself. This could mean deciding that teams only get one development environment - which simply requires a quota change.

It's important to challenge product teams to look for efficiencies in their use of the configuration management product and share their results with each other. Gamification can be used to further incentivise this activity. Firms can target a single product and ask the owners of the configuration management solution to assist them directly. This can all be achieved through the usual IT leadership status meeting, with no need for direct heavy-handed involvement.

By doing this, prioritisation of day-to-day activities is still devolved to the project management function, but there is now no fear of conflicts. Rather than having to step in personally to resolve priority conflicts between teams, the IT leader need only ensure their direct reports are aware of any urgent priorities and let them re-orientate their teams accordingly.

When situations change unpredictably, such as services failing to scale linearly and hitting a wall, the IT leader isn't forced to disrupt entire teams by pulling software, test and or operations engineers away from their current priorities. The team that owns the failing product is right there with all the necessary skills required to troubleshoot and mitigate the incident as soon as it occurs.

Conclusion

In the 21st century, everything built in IT is an online service. Customers expect a business to acknowledge problems with services immediately and take action.

The infrastructure that hosts products, as well as the tests that assure its quality and capability, are created by applications. These infrastructure and test applications are products in their own right and stand next to a host of other solutions required by the IT organisation.

The old organisational structures that many have been labouring under for the past 60 years have become less relevant for what businesses are trying to achieve today. In fact, these structures have now become obstructions.

As this chapter has shown, a shift away from archaic structures, towards a product-centric view of the world, can free firms from old conflicts and restrictions. Product-teams remove the need for negotiation, making it possible to prioritise work in the context of the overall business objectives.

Finally, there needn't be a role for bribery in IT. When teams have the right mix of skills and are focussed on the same goal, nothing can stop them.

CHAPTER SIX

Strategy, tactics and planning, planning, tactics and strategy

In this chapter…

We will see what constitutes a good IT strategy and how this can benefit the organisation as a whole. We will also examine the effects of a great strategy on an organisation's engineers, and their leaders.

We'll also look at a framework that IT organisations can use to enable successful execution of medium-term goals, leading directly to the achievement of strategic objectives.

Finally, we'll see why IT strategy is even more important when leading teams using DevOps methods.

Definition of terms

Before discussing strategy in detail, it's important to be clear about what it means. The words 'strategy', 'tactics' and 'plan' are often bandied about. But when someone says they have a 'sound strategy', does it explain how their goals will be achieved? What does the phrase, 'tactical errors' mean: were the tactics actually formulated and designed to achieve strategic goals? When circumstances conspire against the 'plan', was there actually a plan - or just a vague hope?

A strategy is a statement that describes how goals will be achieved and success will be determined. It should, for clarity's sake, include a statement about what those goals are, but the important word is 'how'.

The word 'strategy' comes from the Greek 'stratēgia', meaning 'generalship'. This implies it should come from someone whose primary concerns are the overall objectives.

Meanwhile, 'tactics' describe the interactions of people and resources to achieve specific near-term objectives. The word comes form the Greek 'taktikē' (tekhnē), meaning 'ordered' or 'arranged'. Whether on the battlefield or in the boardroom, the idea is that tactics describe how people and resources are deployed.

The successful achievement of all tactical objectives should lead directly to the attainment of strategic goals. Think of tactical objectives as milestones on the road to these.

'Plans' describe the individual tasks; the order in which they should be executed; and the interdependencies between them. 'Plans' also describe in detail the exact steps required to achieve tactical objectives.

These definitions are aligned to the roles in the IT organisation. It's clear that only a leader with a good understanding of the business objectives and capabilities and limitations of IT can formulate a good strategy.

As such, creating tactics to achieve strategic goals should be the remit of those with a detailed knowledge of the capability of the people and disposition of resources in the organisation.

Finally, the detail for plans should come from those people who understand the actual specific tasks required. They should be compiled and ordered by someone with the remit to manage interdependencies between other plans across the business.

The benefits of a good IT strategy

A good IT strategy helps the people in the organisation to create a personal connection to business goals.

People are compelled to form tribes. At school, children form gangs or join clubs. Then at university, students seek groups with common interests such as for sports or social activities. Tribes are seen every day in the form of cliques at work and even online forums.

Firms can use this to their advantage. Engineers, managers and leaders will form a tribe around a compelling IT strategy if presented with the opportunity to do so.

Without a statement of IT strategy, the people in the organisation have no idea what they need to do to help achieve business goals. How do people with software engineering skills increase market share? How do test engineers increase average revenue per user (ARPU)? While operations can provide hundreds of ways to reduce costs, turning off all the servers won't really help.

It is far more constructive for business to tell them:

"We're going to increase our market share by building higher quality software. That will be achieved using test-driven development, coupled with behaviour-driven development tools."

This means everyone will understand exactly how they can contribute to increasing market share. A great IT strategy gives employees an insight into how the world will be different once they have achieved their strategic objectives.

The example above can also be added to:

"From now on, we'll receive all our requirements in the form of tests: these will inform our unit and be used as functional tests. When we find a bug, we'll code tests for it and add these into the process, stopping the problem from ever coming back again."

This gives people a vision to aim for; it gets them excited because they can see how this will improve their own quality of life - as well as helping them contribute to achieving the goals of the business.

IT strategy must be informed by a company's objectives. After all, it will assist in achieving these. It should be created with an awareness of the market in which the business operates and take into account the regulations it must abide by. It must also be aware of the resources and capabilities available to the business. It must strike the right balance between the state-of-the-art and the tried-and-tested, across the whole spectrum of disciplines in IT. An IT strategy that does this will also inspire confidence in the rest of the business, its markets, and its customers.

Yet individuals capable of creating a concise statement with this sort of scope are incredibly rare. Crafting a great IT strategy is more usually a team effort.

I've been on many strategy away-days in my time. These have seen me sitting beside country club pools, drinking brandy and smoking cigars. I've taken a Segway tour around San Francisco Bay.

I've watched the Washington Caps beat the New York Rangers from the owner's box. I've worked with some amazing people and created some great strategies. But none of the organisations I've worked for have then put the necessary measures in place to see those strategies executed successfully.

Strategy must be targeted

One of the hallmarks of a great strategy is that it's well communicated. No matter how sound the principles are, if they are poorly communicated, they will have a completely opposite effect.

Good communication is targeted. Creating a 3,000 word 'strategy' document clearly aimed at stock market analysts and mailing it to your whole company is a lousy way to communicate with them.

The advent of online professional networks and technology-focused venture capitalists has seen senior IT leaders take on a very public persona. On the plus side, a career in technology is no longer sneered at as the domain of smelly, bearded geeks wearing heavy metal t-shirts and combat trousers. Those smelly, bearded geeks are still here; they just earn a lot more money these days.

On the negative side, PR aimed at market analysts, investors and partners is often disguised as IT strategy. This does a gross disservice to organisations.

IT strategy should be aimed squarely at the organisation, uniting engineers behind a common purpose. The IT strategy of large firms is a matter for public record and if market analysts need translation, it can be provided. If others in the business don't understand the strategy, workshops can be provided. The strategy should not be diluted or simplified, or it risks turning its most important audience against it.

The role of tactics

Having a great IT strategy that is communicated well gives everyone a solid connection to business objectives. It even shows them what they can look forward to, should they be successful.

What a great strategy won't do, is tell people what they need to do now and in the coming days, weeks and months. It won't provide something tangible to focus on immediately. Nor will it set employees back on course should something unexpected happen, or give them the resources they require to be successful.

Tactics describe resource allocation, capability and the focus of the people in the organisation. They set expectations by establishing what an employees' normal day will look like.

Tactics create a working environment where everyone knows what's expected of them. They establish a normalcy that can be returned to when there's a major event or crisis. Great tactics ensure everyone's efforts combine to form a coherent push towards the strategic objectives. They also align priorities.

Tactics are the way in which an organisation can keep track of what its people are working on; see how much capability it has; and how this is being deployed.

If the strategy can be said to describe what the organisation is going to achieve and how, the tactics denote who will be involved and where their efforts will be deployed.

Good strategy should come from a senior leadership with a broad awareness and knowledge of the business, the market, and the capabilities of the industry. Great tactics need to be devised by a leadership aware of how resources are deployed. Tactics need to be created by a group who know the focus of the engineers, and have the ability to identify the problems that can impede their progress.

A promising strategy is seen as a victory in its own right because it forces those mired in resource deployment, risk mitigation, and prioritisation to shift their perspective away from the details, towards the wider world around them. Leaders at this level contribute their best if, having helped formulate the strategy, they can work with their peers and bosses to create the tactics necessary to achieve the strategic goals.

This is where everything goes wrong in traditionally structured organisations. The people at this level are forced to work against each other by the organisational structures they operate within.

Software engineering leaders are pushing for constant change and project progress. The leaders of operations-type functions try to limit change and maintain stability. At the same time, those in charge of testing are tearing their teams apart trying to move people to projects just as they need them, while also trying to staff automated functions.

Called in to address the results of these conflicts, leadership focuses on the software build and deployment process. After all, software engineering is perceived as the function that creates the revenue-generating opportunities. It's the largest group of people in the organisation and its high costs are offset, to a certain extent, by capitalisation of the products they create - and R&D tax relief.

The theory goes, that if everyone else in the IT organisation is aligned with the software engineering function, priority conflicts must be resolved. Unfortunately, this idea doesn't consider the essential functions that are not performed by software engineering. This happens because senior leadership, having received an escalation, is forced to react to long-running organisational problems in the moment.

This focus on software engineering leads to a blinkered approach and actually causes more problems than it solves. It also prevents the business from realising more opportunities for capitalisation, tax relief and positive PR.

All development work can be capitalised as an asset - and not just that which directly generates revenue. It also includes work undertaken to create automated test suites, configuration management, system integration, and monitoring solutions. This is categorised as 'application and infrastructure costs', so it can be capitalised in exactly the same way as the direct revenue-generating products. This gives firms yet another reason to create product-focused teams rather than skill set silos.

By considering all development work together, whether revenue generating or revenue supporting, it becomes liable for R&D tax relief. This is particularly easy to justify for automated testing and DevOps, as these two disciplines are so nascent. The applications that make up the automated test, infrastructure build and support work can be published as open source. As the intellectual property involved is not directly revenue generating, the business loses nothing. The argument to tax authorities is then even easier to make.

Instead of the traditional IT structure where teams are categorised and segregated by skills, people can be focused on a set of products. Consider, as suggested in the previous chapter, that each team owns a set of products, some of which might directly generate revenue and others that can support this.

By considering this in their IT strategy, leaders will not be in direct opposition to each other's goals. Instead, their relationships are now defined by the products they are responsible for. When these people get around a table and discuss how best to deploy their team's capabilities and resources, they can take into account the organisation's strengths and weaknesses in terms of the product.

This is a lot less emotionally charged than identifying a team or individual's strengths and weaknesses. It leads to a much more honest discussion, allowing people to more readily accept suggestions from their peers. Now it's not operations telling development how to do its job, it's one product team making suggestions to another.

When considering IT strategy, leaders should first identify the products that directly contribute. The initial focus will be the products that generate revenue; grow market share; are responsible for the highest costs; or that will generate the most industry publicity. Having defined the additional capabilities and performance required by these products, attention can be turned to those that support them. Are additional capabilities needed? Are there performance problems that need to be resolved? The leader's direct reports should now be able to address the strategic goals - which they hopefully helped create - with tactical product choices.

Examine once again the example strategy proposed earlier:

"We're going to increase our market share by building higher quality software. That will be achieved using test-driven development, coupled with behaviour-driven development tools. From now on, we'll receive all our requirements in the form of tests: these will inform our unit and be used as functional tests. When we find a bug, we'll code tests for it and add these into the process, stopping the problem from ever coming back again."

This strategy requires a variety of different products and development projects. A test suite will be needed and the team will have to code a bunch of automated tests.

In order to be able to troubleshoot a problem in a live environment; define a test case; fix the bug; and retest quickly, configuration management and system integration tools are needed that allow the building of multiple environments, quickly. That's going to require some sort of virtualisation toolset; perhaps a cloud solution.

Working from the strategy, those that report directly to the leader now have to choose which solutions to buy and which need to be built. They need to decide how those solutions need to integrate and how teams will work with them. They will examine existing capabilities in the organisation and identify gaps and choose where best to deploy engineering time, money and resources.

By following through on these decisions with plans and using those plans to align the teams a firm will be well on its way to meeting its strategic goals.

So a successful IT organisation establishes the deployment of its people, their focus, and the distribution of its resources by the product. Each product should receive staffing, resources and requirements appropriate to its role in the attainment of strategic objectives.

The prioritisation of work is defined by the product's contribution to the overall strategic goals. Certain products will require features before others, while some capabilities need to be available at particular times and possibly only during specific events. These fixed points and times are the milestones of product plans.

Finally, the products should define the team's working day. Some products require much more development than support, and others are the opposite. At the same time, some products require minimal security, with others needing intense active scrutiny. These differing needs should provide the requirements for the team who will work on them. A wise leader will chose appropriate people by first understanding the capabilities and temperament that will best serve the product.

The place for planning

Once a strategy has been communicated and tactical approaches designed, organisations can begin planning.

Planning in this environment follows the same process as always: goals and deadlines are defined; product requirements gathered; experts consulted and estimates charted. What will change is the execution of those plans.

Agile development orders us to respect the chaotic nature of life and reign ambitions in. By focusing on the capability available in the coming weeks, the impact of changing requirements can be managed.

Scrum masters were added to the project management structure to negotiate with stakeholders in order to meet changing requirements and keep software engineers focussed. Agile development has been amazingly successful in this regard. So successful, in fact, that I haven't worked in any other environment in over a decade.

When agile development was conceived, IT worked under the assumption that software projects ended at some point. The project team would celebrate their achievements and go their separate ways once requirements had been developed.

A decade on, this is no longer the case. Products aren't applications that are shipped and sold; they are services. Services have subscribers; they support in-app purchasing and advertising. They have lifespans measured in years and are under constant development. And the service that stops developing, dies.

Agile development will get services to their launch date, but everything goes wrong as soon as service ownership moves outside of the software engineering function. However, by extending agile development with DevOps - or better still, a product-focused approach - services can grow, develop and succeed without interruption.

This means that the project management structure takes into account not just the requirements as defined by business stakeholders, but also the team who will support the product in the live environment over the coming years. The team responsible for the performance, security, data integrity, infrastructure costs and quality assurance of the product is the same group that is building it in the first place. Now, the natural optimism of software engineers will be tempered by the pragmatism of operations and the thorough attitude of testers.

Plans no longer perish at launch; they never end. As long as the product lives, there are plans. By following this structure, people in the team and resources in use can be accounted for and managed throughout the product lifecycle.

Sensors, logic gates and actuators

If the universe was well-ordered and predictable, IT leadership could now sit back and relax, safe in the knowledge that an excellent strategy had birthed equally fine tactics - and these carefully crafted plans would proceed through to a successful conclusion.

But the universe isn't so. Markets change; new technologies revolutionise what's possible. People have babies, they get better offers, they get sick. Businesses are bought, sold and merged. But throughout this, products need to generate revenue and excite new customers while pleasing existing ones - and still get rave reviews in the media.

In order to remain successful, businesses must recognise opportunities and mitigate risks as soon as they arise. This means ensuring visibility, decision-making and control are at the right levels. Visibility must inform decision-making, which must, in turn, inform control.

However, many modern businesses struggle to deliver strategic goals because their structure only allows them to have one feedback loop in place.

Consider that, within most traditionally structured organisations, the lion's share of focus is on the software engineering function. Internally, this team can suffer no problems other than late delivery: all it is tasked with is delivering a piece of software. However, when that software fails to deploy properly or suffers problems in the live environment not encountered in development or testing, the software engineering team is powerless on its own.

At this point, software engineering needs to enlist the help of other teams. If priority conflicts occur, the only route available is through escalation to the leadership. Of course, the leadership must act and so day-to-day activities can be disrupted. Teams are forced into crisis mode: stress levels rise, introverts begin shouting, while extroverts sulk. It takes days - or weeks - for the organisation to return to a normal equilibrium and that's assuming no further crises occur.

These situations create undesirable patterns and see organisations falling into crisis after crisis. For a while, the business will fool itself into thinking that crisis management is a strength and the constant hustle and bustle is a sign of progress. However, a dispassionate observer would note how few technical innovations occur during this time and how little progress is made towards strategic goals.

And as this situation demonstrates, it's not enough to have methods for creating and communicating strategy; generating tactical approaches; creating plans; and building products. The product lifecycle events must feed back into plans - and planning process should input into tactics. At the same time, tactics need to initiate a strategic review when circumstances require it.

Modern planning processes and tools have excellent feedback mechanisms built into them already. The whole planning process generates anecdotal feedback. Combined with narrative from the project management structure, task progress tracked on a burn down chart creates truly useful information that can be used for decision-making.

Providing the product team has the skills to build, launch, manage and support the product, feedback can be used in real-time. This team's focus is on the product, rather than the software development lifecycle: people will not wait until they are "dev complete" before deploying the application into a system integration environment. There are also the right perspectives in the team at the planning stage and during the sprint to interpret task progress correctly and take action accordingly.

If, for example, the team discovers that what at first appeared to be a simple requirement suddenly needs additional infrastructure, priority decisions can be made. The team can also choose to negotiate with product management to delay the delivery of that requirement or propose that other requirements be dropped from this sprint to make room for additional infrastructure and testing. If all the requirements are essential, the team can assess whether bringing in additional people or working more hours is the right solution.

What's important is that this assessment happens instantly. I've lost count of the number of crises that occur at 16:00 on a Friday; I can't recall a single problem that kicked off at noon on a Wednesday.

But some problems are bigger than a single product team. If an IT organisation is to successfully deliver, week in and week out, regardless of the situations it's faced with, there also needs to be a feedback loop, from the product lifecycle into the company's tactics.

The product team's day-to-day experiences should inform the tactical approach. The planning function is ideally placed to provide this feedback.

If one or more of the product teams isn't getting what it needs from the others, a tactical problem needs to be investigated. It could be that one particular product requires more from another team than it has to give. That team might need more people; focus could be off; or the team might need access to unavailable resources.

The solution could be trivial: it might be a matter of the team re-scheduling some work. It could also be more serious and indicate a lack of the right people or resources. This problem might not be solvable at the tactical level and could necessitate a strategic review. The important thing is, that these problems are raised in the normal course of events as just another management problem.

The leaders of the product teams can present the situation to their peers and leadership for discussion. Now, rather than dragging senior leaders straight into day-to-day task level activities where their decisions can have a raft of unintended consequences, the group can discuss how the problem should be solved: tactically or strategically. The answer could be to relax the standards defined in the strategy for just one product. Or it might be enough to temporarily reassign some people in order to help a product through a tough patch.

Opponents of this type of approach argue that this impedes decisive action. They emphasise that in today's modern business, decisions need to be made and executed quickly. I would argue that frequent, rapid-fire decision making from the senior leadership level into an organisation's day-to-day tasks is indicative of a business hurtling out of control.

A place for everything...

The rewards that come from creating appropriate structures for strategy, tactics and planning - as well as the feedback loops between them - are manifold.

Senior leadership will be better able to focus on the future without the constant distractions of the now. These kinds of structures prepare organisations for market and technology changes, removing much of the pain. This, in turn, makes an organisation better able to cope with change itself.

As mentioned in Chapter 4, large organisations around the world struggle with communications. Employees don't know what's going on and ask for more. They want to know how they're doing, as well as the impact of their contribution and that of their department. Ultimately, this means knowing that they're doing the right thing.

Communicating strategy, tactics and planning clearly so that employees' day-to-day tasks relate directly to strategic goals solves that problem. Any communication about market or industry changes, regulation, acquisitions or mergers can be phrased using these mechanisms. This allows everyone to understand the impact of their contribution and receive feedback in almost real-time.

Any required change in priorities is described at the strategic level and informs the tactics and, as a result, plans change. An entire organisation can pivot in a day without any drama or crisis. Communication occurs appropriately from senior leadership, directors, line-managers and the project management structure.

This leads directly to the culture so many organisations crave. How many hours have been spent discussing how to achieve or retain dynamic responsible cultures?

Providing that the leadership clearly expresses its goals, creates the structures necessary to achieve them, and also allows employees to contribute directly and receive real-time feedback, firms can enjoy the culture they're looking for.

This is exactly what happens within small start-ups. In many ways, it is what is natural for this type of business: a small group of people obsessively focused on a common goal have these traits. What has proven difficult, until now, is bringing these behaviours into larger groups.

This first failure is obvious: the goal is wrong. Large groups of people can't be treated in the same way as small teams or individuals. The psychology of crowds is not well understood and not at all predictable. There are much more sophisticated models to describe individuals and small groups. As most people experience them daily, they are naturally better able to deal with individuals and small groups than crowds.

The key to retaining or creating a focused and responsible culture of valuable contribution is to keep groups small.

DevOps: the double-edged sword

DevOps brings many advantages: the published evidence is just coming to light now. It seems fairly conclusive that organisations following DevOps principles are deploying code more often and suffering less live service incidents per release. From my own experience, DevOps creates a much better working environment, bringing together all the IT disciplines while focusing people on the business objectives. DevOps, once established, also brings incredibly rapid innovation. I have carefully avoided using the word 'progress' here for a reason.

DevOps allows teams to very quickly try out solutions and integrate them into software build and deploy functions. But as with all innovation, a good many of these solutions solve one problem at the expense of others.

The ability to rapidly work through these problems eventually leads to innovative solutions - yet the journey is not a straight-line. Judged at any given point in time, it can appear as if a lot of effort has resulted in very little real-world gain. It's not at all unusual to be forced to pivot part of the way though a project or drop a tool previously considered essential.

A clearly-defined strategy, accompanied by coordinated tactics and plans, provides the groups that are innovating with the guidance they need to focus their creativity. It removes the risk of teams losing their focus and re-formulating the problems as a result delivering solutions that fail to solve the issues the organisation faces.

As businesses grow, it's not unusual to find engineers hired to support specific endeavours. These people will invest heavily in a tool or particular method because it solves the problem they're facing at that time.

Later, when the team and remit grows, several different tools and technologies will be operating in the same functional space - and the team will find itself context-switching frequently between these. But there will come a day when the team needs to consolidate these tools.

There will then follow a period of investigation and testing. Hundreds of person hours will result in approaches debated, tools tested and new capabilities developed. Eventually, a solution will improve the lives of all involved, allowing the organisation to thrive - but at a huge cost in person-hours. This situation isn't inevitable.

If a business is going to start small; grow as it needs to; and run multiple, simultaneous and independent projects overseen by different leaders, a single set of guiding strategic objectives and tactics brings the benefits of a shared approach. If it's the intention of the organisation to introduce projects that test new approaches, the scope of these tests will be clearly defined and accounted for. The cost of reintegrating these new approaches or running them for the long term will be taken into account.

Without the guiding hands of strategy and tactics, the gains possible from hundreds of potentially productive hours of DevOps work are being lost. The individuals involved are certainly learning - and adding keywords to their CV - but the wider business and IT organisation is only benefiting marginally. Only if the organisation has clear, strategic objectives and tactical approaches can the vast potential of DevOps be truly realised.

Budget levers

Every organisation - large or small, start-up, mature, or multinational global giant - needs to change its budget allocations at various times throughout its life. For the IT organisation, this is inevitably the time when projects are culled, hasty investigations into infrastructure costs begin and department heads enter endless rounds of budget submissions.

This usually sees the introduction of complex Excel spreadsheets categorising costs against projects. The testing teams are asked to calculate how many hours each engineer spent on individual projects. Frustrated that each operations engineer works on half a dozen projects every day - half of which don't even have budgets - they are categorised as 'general cost', or the expense is spread across all of them.

Contrast that with the product team organisation where all effort is put towards budgeted products. Products such as Configuration Management, Network Security and Continuous

Integration have their own budget allocations. Firms can now take that concept a step further and give the teams owning these budget responsibility. Now, when the organisation needs to change, the budget allocations are product decisions. For example:

- Which product will receive less development over the coming couple of quarters?
- What compromises can the organisation accept?
- Where are there opportunities to reschedule work to push requirements into the next budget allocation period?
- Is each product team efficient?
- Is there scope for reducing one product team's budget allocation but making up the shortfall with infrastructure efficiency gains?

Now, rather than embroiling senior leadership in the minutiae of day-to-day departmental business, the leadership can take a product view and challenge its direct reports to look for efficiencies in the teams. The product teams can be challenged to respond to the call for efficiency by making recommendations of their own.

The feedbacks loops created between planning, tactical approaches and strategy can be used to get everyone at each level working constructively towards the new budget goals.

When strategy receives new information, it can initially decide that it wants to stay its course but look for efficiencies. Strategy could challenge tactics to create efficiencies or innovate. In turn, tactics might challenge the planning function to reschedule.

The product teams feed back, through planning to tactics, and into strategy. This way changes can be introduced in a controlled fashion at every level. At the same time, each level is informed by the one above and below it. This way, there is no need for budget changes to become a crisis.

Conclusion

With the growth in capability presented by DevOps or a product-centric approach, IT leadership needs to grow its own ability. Progress is not the same as activity: skilled and talented people need strategic leadership, backed by sound tactical approaches and planning to turn activity into progress.

This growth in leadership capability brings with it ever-increasing rewards: more control and oversight is available to leaders. Change is accommodated easily and with less drama. Response to change accelerates and that, in turn, speeds up progress further.

With increased control and oversight, authority can be safely devolved and this too accelerates the ability for organisations to progress and respond to change.

CHAPTER SEVEN

The DevOps Virtuous Circle

In this chapter…

We'll examine how testing unlocks DevOps' potential. As this chapter will demonstrate, testing can provide the product team with focus. That's of course after the strategy has provided direction and the tactics have made the necessary tools and resources available.

We'll look at a series of steps that when taken in order, lead directly to continuous integration and closer to a process of continuous improvement.

The most common mistake

The most common mistake IT organisations make is thinking that testing takes place once the product or feature has been built. This attitude limits testing to poking at a product from the outside. It's like taking a car for a short test drive to determine its suitability for a 24-hour endurance race. Not understanding the design decisions made during the build processes or potential weaknesses these decisions might have engendered - leaves the 'test drive' approach wanting.

Test driven development (TDD) has been proposed by some as a response to this. The theory behind TDD is simple and compelling: having received some reasonably-specified requirements - ideally in the form of use cases - the software engineer writes a test. This is intended to verify that the software meets the requirements once integrated into the build and, at first, it should fail. This demonstrates that the build process and test harness works and confirms that no-one else has tackled these requirements.

The software engineer then writes a reasonable minimum amount of code to pass the tests. Once the tests are passed, the code is re-factored, cleaned, duplication is removed and variable and method names are checked to make sure they adequately reflect usage. Finally, the test is executed again and should pass, running against good clean code.

TDD's usefulness is still debated by the people who created it, and many software engineers struggle to identify when it is appropriate. However, it is very clear is that no-one ever intended it to be a quality assurance technique. In fact, it's widely agreed that TDD isn't actually a test mechanism at all.

TDD can be a useful way to specify the task that code has to perform; it can provide documentation for code. But TDD is definitely not a reliable practice when it comes to assuring the quality of a product.

BDD to the rescue!

Developed from TDD, behaviour driven development (BDD) takes a slightly different approach. It recommends that software be described in terms of its desired behaviour, using the form:

GIVEN some initial starting condition
AND some additional other condition
WHEN an action is performed
THEN some behavioural results

Like TDD, BDD isn't really a testing solution either. It is a framework that teams can use to help understand requirements. While BDD tests give software engineers assurance that the code they have created exhibits desired behaviours, it does not tell them about undesirable behaviours. However, BDD does boast some qualities that TDD doesn't.

BDD requires that the organisation chooses - or builds - a tool to execute tests and creates definitions for each step described in the behaviours. Once this is done, behaviours can be tested every time the software builds. Because behaviours can be read and understood by anyone in the business, it's very easy to display the progress of a software build and explain exactly which behaviours have been verified and which have not.

By displaying, on a dashboard, which tests have passed and which remain IT organisations can give everyone in the business visibility into progress and as a result, create a greater sense of confidence in the IT organisation.

This is all pretty good, but if BDD isn't primarily a testing mechanism, how can it help to assure the quality of our products?

Product Lifecycle not just SDLC

The key to unlocking the potential of BDD for testing is putting it to use by a product team: these are people with software, operations and test engineering skills focused not on their individual disciplines, but on their products. BDD can focus these people, however disparate their capabilities, knowledge, experiences and outlooks, on the behaviour of the products.

This will see those with software engineering expertise initially focusing on how code behaves GIVEN initial predefined conditions; AND some defined inputs; WHEN an action is performed.

The people with operations engineering expertise will focus on how the application and the systems behave GIVEN initial predefined conditions; AND when given some defined inputs; WHEN actions are performed.

The test engineering experts will focus on how the application behaves GIVEN initial predefined conditions; AND when given some defined inputs; WHEN actions are performed.

However, when the group discusses behaviours together, it will begin talking and thinking as a team. When analysing new requirements or discussing challenges, the framework BDD provides will encourage a focus on the product's behaviours end-to-end. This is not the same as the code's behaviour, nor that of the system, or the application. The focus of this team is finding a shared understanding of how the application behaves on its infrastructure when it is used.

The team, using BDD to share individual perspectives, will look at the product from both the inside and outside and in all the contexts in which it has to perform. The group can then share perspectives and determine an appropriate suite of tests that are needed to verify the behaviours of the product.

Together, the team can make pragmatic decisions about exactly how much testing can be automated and where it should be manual. They can work together to ensure the build pipeline interfaces properly with the configuration management system and the test tools, ensuring that products are tested on an appropriate range of environments. The team can also make informed decisions about how far system integration testing needs to go.

Informed by experience, the team can decide when it's appropriate to perform full system integration testing - and it'll understand the ramifications of that decision. The teams can now choose whether it would be more appropriate to code a mock; just use a stub; or test against a full version of a particular service informed by all of their experience and perspectives.

Exploratory testing AND automated testing

Even when used by a product-focused team, BDD will only guarantee that the behaviours it tests for are present: it won't provide assurance that undesirable behaviours or defects are absent. Nothing can guarantee a product is free from defects; the goal of software testing is to reduce the risk of their occurrence.

Combining exploratory testing with automated testing minimises the risk of defects and almost eliminates the likelihood of these returning. This only works if a process like BDD is followed, with desirable behaviours tested for using an automated test suite. Without this, the team has to spend too many hours verifying things that could be tested perfectly well by systems.

Freed from the need to slavishly verify every desired behaviour, the product team can use its knowledge of the systems already tested to explore. These well-informed engineers are looking to find weaknesses in the systems they built. They won't take an optimistic approach: that is taken care of by the automated test suite, which has already confirmed, or is in the process of confirming that the product behaves as specified.

When exploration finds defects or aberrant behaviours, the team will code tests to detect them, before coding the fixes. This is then added to the test suite, executing alongside other tests against the build.

By using the automated suite to execute every test that finds a defect, errors can never re-occur. Each time the team discovers a problem, the test can be duplicated. This means the defect will be tested for every time the software builds.

Each subsequent build will then be tested for all the desired behaviours, including those previously specified - and for any defects discovered during exploration. In other words, the team maximises the time it spends testing new behaviours and conditions, at the same time minimising the amount of repeat work required.

This has a huge impact, not just on the quality of the software, but also on morale. Engineers hate repetitive tasks: that's what computers are for.

Design for testing

Products built without a focus on verifying behaviour are often extremely difficult to test. Once a product has reached this point, the only solution is to throw people at it. This is expensive: forcing people to repeat the same test script every day will not yield value for money.

I experienced this issue while working for Playfish. At the time, Playfish games were comprised of a Flash client - which constituted the game as far as the player was concerned - and a server-side Java application, which validated user inputs and recorded progress.

The player's game session was protected by an authorisation and authentication mechanism that meant no-one could interfere once it had begun. This ensured players' progress was protected and billing transactions could begin securely.

But it was this mechanism that made it so difficult for Playfish to perform automated, load and performance testing. In theory, it should have been a simple matter to make requests to the server-side component and drive a simulated game session. This would have enabled us to automate a huge range of test cases, as well as allowing us to benchmark performance at build time and load test games effectively prior to release.

The security mechanism was never designed to allow for testing, which meant that the only automated testing we could initially perform was using front-end client mechanisms such as Automated UI tester for Actionscript.

Unfortunately, this mechanism was invalidated every time the game interface changed. The game was displayed in a browser, on Facebook, within a console - and subject to modifications by marketing and payments teams - so such change was an almost weekly occurrence. Even if we had been able to use these mechanisms, they are slow to execute because interactions occur in real-time and so can't practically scale for load testing.

If an organisation designs its products to be tested, the act of testing itself becomes much easier and quicker. If someone at Playfish had designed a mechanism that allowed game playing to be simulated on the servers, basic functions could have been executed at build time very rapidly. This would have saved thousands of hours of test engineers' time.

If that mechanism was kept up to date by each game team and timings were added, performance testing could have occurred after every build. Load testing would have been a simple matter of modelling and creating a few scripts. That mechanism would have paid Playfish back every day. The return on investment of just a simple tool would have been enormous.

And it isn't just software that needs to be designed to be tested, systems require this too.

The role of Configuration Management in the Development process

Configuration management describes the activities required to define system configurations and build solutions using these. It is the product in which system arrangements are defined - and this initiates system build activities.

Configuration management can build one system, or many. When it is used to build multiple systems with varying interdependencies, it is called 'orchestration'. These two concepts are the next great enablers that allow the team to test its product.

To many in IT, configuration management is a tool for locking infrastructure into single known configurations. Internally at AOL and Electronic Arts, it was just a tool to build lots of infrastructure quickly and make sure each server was the same as the next.

But during my time at Playfish, we extended the capabilities of configuration management further, adding orchestration so entire products could be built and deployed.

British Gas Connected Homes had an even more enlightened view of configuration management. This allows it to deploy product upgrades seamlessly into the live environment. It can be even more than that, though.

As with BDD, configuration management doesn't need tools in order to be valuable. However, like BDD, the right tools can save considerable time and effort.

Essentially, configuration management systems allow teams to define the infrastructure for their products as code. Having created the software, the build system can instruct the configuration management solution to build infrastructure and deploy that software.

Take this concept one step further and builds can be tagged as functional test builds and infrastructure created with suitable mocks and stubs for such purposes.

If all the necessary products have been built to enable testing, they can be deployed on system integration infrastructure, created by configuration management systems, and integration testing can be performed. Databases can be built with test data, properties configured appropriately, and automated tests executed.

If the progress of each of these processes is exposed on dashboards, the entire business knows the status of every product. If some products are nearing release, multiple system integration environments can be built containing several versions. This ensures compatibility and keeps releases independent.

If every product in the stack requires its own infrastructure for testing, it is going to get expensive very quickly. This is another reason why products should be designed to be tested: it ensures that they can reside on minimal, yet still representative, infrastructure.

This was another problem we faced at Playfish: because the services had not been designed to be tested, it was practically impossible to make them reside on the same infrastructure. Therefore, a full system integration environment required 16 different pieces of infrastructure. It became a very expensive activity.

Clouds for everyone!

There's a lot of noise surrounding cloud computing. Some criticise it for being too expensive, with others saying the tool ends up being cheaper when considering the total cost of ownership (TCO). It's therefore very difficult to get a conclusive answer.

However, there is one definitive use case where cloud solutions are much cheaper and enable greater productivity: testing.

Cloud computing costs nothing when there is no infrastructure running. So, if products have been designed to be tested and configuration management systems are in place, test infrastructure can be created when needed and decommissioned when it's not.

From any practical standpoint, test infrastructure is infinite and available with no risk of disrupting another project's work. Not everyone will agree; some will argue: "Teams find it very easy to acquire cloud infrastructure, but they never shut it down. It ends up costing more than it would have done if they had simply purchased hardware."

This is true for many organisations. However, this problem doesn't occur if product teams manage their own budgets and are held accountable for costs.

Dashboards

Within larger businesses, product teams don't operate in isolation. Products need to work together; front-end services require back-end services. At the same time, build systems rely on artefact repositories and third party-provided libraries. In addition, these elements rely on a wide variety of ancillary solutions such as content delivery networks (CDNs), domain-name services (DNS) – and even network connections.

Each of these services can reside in one of at least six possible states: available or unavailable - or flapping (rapidly shifting states from available to unavailable, often caused by mis-configured failure mitigation mechanisms such as failover systems). More interestingly, some systems are about to be updated; they could have just been updated, or they could have been compromised.

These states should be available for product teams to view on dashboards. Dashboards force teams to expose their state in a standard, easily consumable form. This information should include the version number of the current release, as well as that of the expected next one along with its estimated launch date. This information helps product teams to determine which versions to use for integration testing.

Dashboards are a product and considering them enables all kinds of possibilities that aren't immediately obvious. They are useful for everyone in the business, but it's still rare for organisations to invest in dashboards. This is another example of how firms lose out by failing to consider everything built or used as a product.

And there is more to dashboards than product development: they can display development progress such as story states; the progress of automated tests; test coverage metrics; test environment configuration and performance metrics; and how long the build process takes.

Dashboards also provide a mechanism for displaying a service status to customers and partners. Once a dashboard product exists, it can be used to display any information at all. For example, the tool can be context sensitive and display sales, engagement, or revenue.

During my time at Playfish, we created a dashboard displaying game release deployment information. It listed the game name; the client and server release numbers; date and time of deployment; a link to the release notes; and the name of the engineer that performed the deployment.

The data was automatically populated by the deployment tool. This was useful for everyone as it allowed teams to get an instant update on the state of game - and service releases - across the estate.

But we didn't expect what happened next: games teams began using the data to challenge each other to produce the lowest number of bug fix releases after each feature was launched.

This pattern was repeated multiple times: when the operations team produced a dashboard, it ended up being used for more purposes than originally planned, in turn adding unexpected value.

One of the first dashboards we created displayed obfuscated information about payment transactions. The ID of each player was obscured, with the size of the game payment transaction, the currency, and the payment provider displayed. This was in the days before Facebook credits, so we had a dozen separate payment partners for different regions.

The dashboard was originally created to help track our progress towards migrating to Facebook credits. Once that project was complete, we realised that the frequency of payment transactions gave us a good indicator of non-critical game problems that weren't picked up by system-level monitoring.

Monitoring: where we've all been going wrong

Often circumstantial and haphazard, monitoring is usually only considered once products have been built. Most people seem to think of it as 'system monitoring', up until the moment when they have a reason to care about application performance and then it's thought of as 'application monitoring'.

This is important because although operating systems provide easy access to a vast wealth of fascinating metrics, they rarely reveal anything interesting about applications. Therefore, when an organisation wants to know if an application is working, it is forced to gather clues like Sherlock Holmes, drawing increasingly accurate deductions by gathering circumstantial evidence.

Non-zero CPU usage means something is happening. Memory utilisation above the configured amount indicates that the application might be running. If it's below the configured amount, it might not be. At the same time, network traffic on the appropriate port could point to real transactions, rather than the usual noise every internet-connected system receives. The CPU isn't waiting on any resources, so presumably the system is performing well. From all this, it can be deciphered that the application should be working well – but that's only an educated guess.

Yet this is so absurdly unnecessary. Applications know whether they are working or not: they know if they are receiving appropriate requests, and if they can access all the resources they need. But because organisations consider monitoring as some sort of after-thought performed by a different team, they are forced to poke at applications with a stick to see if they're alive, rather than processing exceptions directly at the source.

All monitoring systems can be thought of as executing tests: is port 80 responding to requests? Is memory utilisation less than 90%? Is Java present in the process list? Data is gathered, queried and the response checked against a desired behaviour.

Once that's done, the monitoring system decides whether to call the alerting mechanism based on whether the behaviour is within defined parameters. If the result falls outside of the configured range, it will alert.

So monitoring is testing. If the product team is focussed on verifying the behaviour of the product, isn't the final application the monitoring system? Yes! Tests that confirm the behaviour of a product can be executed everywhere: the build and test environments AND the live environment by the monitoring system.

As an interesting aside, all these tests can also cause alerts to be sent. With a little thought, different alerts can be sent for types of failures in test scenarios. Test failures in the live environment might send an SMS and log incidents. Build or integration test failures could log a bug report and change a light from green to red - or cause a small plastic bunny to waggle its ears and speak in an irritating tinny voice.

The Next Gen DevOps Virtuous Circle

The components below form the foundations to allow any IT organisation to rapidly develop, build, test and integrate products.

- A testing framework focussed on behaviour
- An automated test harness
- A strategy that focuses engineers on testing
- Applications designed, from the start, to be testable

• Configuration management solutions that allow the rapid creation of any environment

• Infrastructure that can be rapidly deployed without distraction

• Methods that can deploy products whenever they're ready

• Dashboards displaying the status of the build, all environments and ancillary services

Laying these foundations in precisely the correct manner can create one additional emergent benefit that makes the whole greater than the sum of all of its parts: continuous improvement.

Long considered the holy grail of ITIL, ISO9000 and Six Sigma, continuous improvement is within easy reach of the Next Gen DevOps virtuous circle:

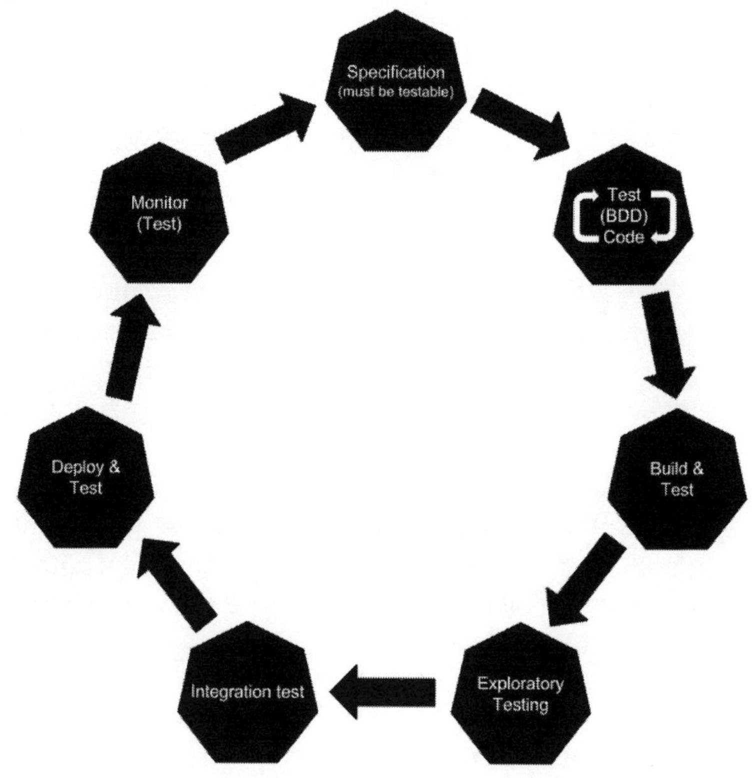

The team must be able to test every requirement it receives. If requirements aren't delivered in a suitable format, the team must interpret them. BDD provides an excellent framework for these conversations with stakeholders.

When the requirement's behaviours have been specified in the GIVE, AND, WHEN, THEN form, they can be analysed by the whole team. The people with software engineering skills will aim to understand the additional capability that's needed and can suggest the tests to verify this.

The people with operations engineering skills will begin considering the impact of the new capability on systems. They will be looking for security, reliability, any essential performance and scaling tests, and any requirement to change infrastructure or configuration.

The test engineers will consider how best to validate the new capability, working out the additional tests necessary and asking whether these can be automated. The team will discuss this as a group and agree which tests are needed, who should create them, and at what stage and in which environment they should be executed.

Finally, they'll discuss if any exploratory testing is required. Potentially, everyone in the team could be given test creation and coding tasks as a result of a single requirement. At this stage, coding of the tests can begin. Tests should be integrated and allowed to fail before the new capability is coded.

Once the new capability has been coded, it can be integrated with all the tests executed against the new build. Builds must complete quickly in order to encourage the team to commit often. Tests that require other services or platforms should use stubs or be mocked as far as possible.

As new capability is successfully added to the build and new tests pass - assuming old ones don't fail - the team can begin exploratory testing. This becomes a vastly more valuable exercise if, once problems are found, tests are coded to find those issues again in the future. These tests can then be added to the suite, with the need to manually probe previously tested capability diminishing every cycle. Eventually, only the new capability and the UI may require any manual testing at all.

At a certain point in the exploratory testing phase, an integration build is prepared. Exploratory testing can begin on the integration build if the team prefers and the environment is conducive to this approach. The important thing is that the entire suite of automated tests that have just been completed are now executed again against a build of the software running in a live-like environment, integrated with live-like services.

The team will have to decide exactly how live-like this integration is. But it will be providing support for the product once it's live, so there will be ample motivation to get this right.

Some exploratory testing will need to take place in the integration environment and this should focus on the interaction of the product with other products and services.

Once testing is complete, the build can be deployed to live. At this point, the software will have been tested three times within three different contexts. First, the behaviour of the code is tested as it's developed. Next, these behaviours are verified again in the build environment with all the code integrated together. Then, the behaviours are verified again, this time in the system integration environment with the build integrated with a mixture of stub, mock, test and live versions of the ancillary services it needs to work with.

As with the coding phase, the deploy phase starts with ensuring appropriate tests are present in the monitoring system. This is where BDD provides the biggest return for the product team: the tests created to verify the behaviour of the product throughout the software development process can now be integrated into the monitoring system.

This fourth round of testing, performed by the monitoring system, verifies the behaviour of the product in the live environment. Results can be directly compared with those obtained from the system integration and build testing as well as the tests performed against individual units of code and modules on the developers' local environment.

If any problems are discovered once the product is live, tests can be created in the test suite and monitoring system, ensuring they don't make their way into the live environment again.

If any changes in the live environment adjust the behaviour of the product, they will be detected by the monitoring system, and the alerts can form new requirements for the team. This way the team is constantly managing and maintaining the behaviour of the product. New tests are added as failure cases are discovered and the quality of the software steadily improves with each release.

Conclusion

By choosing BDD, everyone in the product-focussed IT organisation can describe behaviour using the same language. Only then can a business achieve true transparency, from requirements specification all the way through to product behaviour in the live environment.

By considering the product lifecycle - rather than only the software development lifecycle - these behaviours can be verified both against a software build in a sterile lab environment, and in the real-world, every day, by the monitoring system. The tests aren't just written by software engineers focussed on delivering new capabilities, but also by operations engineers with an intimate knowledge of the systems, and test engineers with expertise in discovering aberrant behaviour. Monitoring is no longer a game of deduction from circumstantial evidence: it's a definitive statement of product behaviour.

Testing unlocks the potential of DevOps, as well as that of the entire IT organisation, through rapid development, testing and deployment of products.

CHAPTER EIGHT

The Product-centric IT Organisation

In this chapter...

We'll examine the organisational structure shared by most IT departments and explore why it's so common.

As this chapter will show, IT organisations often fail to re-organise themselves effectively, and this can obstruct their own progress.

This will lead us to an approach for creating, growing and transitioning towards a product-focussed structure.

The traditional IT organisation

The typical IT organisation follows what's known as a 'functional organisation structure', grouping people with similar skill-sets together. This sees entire teams of operations engineers, testing or QA people; security experts; and service desk workers. Larger departments might have an engineering director tasked with overseeing the activities of several software teams.

Within a medium sized business, the IT organisation might look something like this:

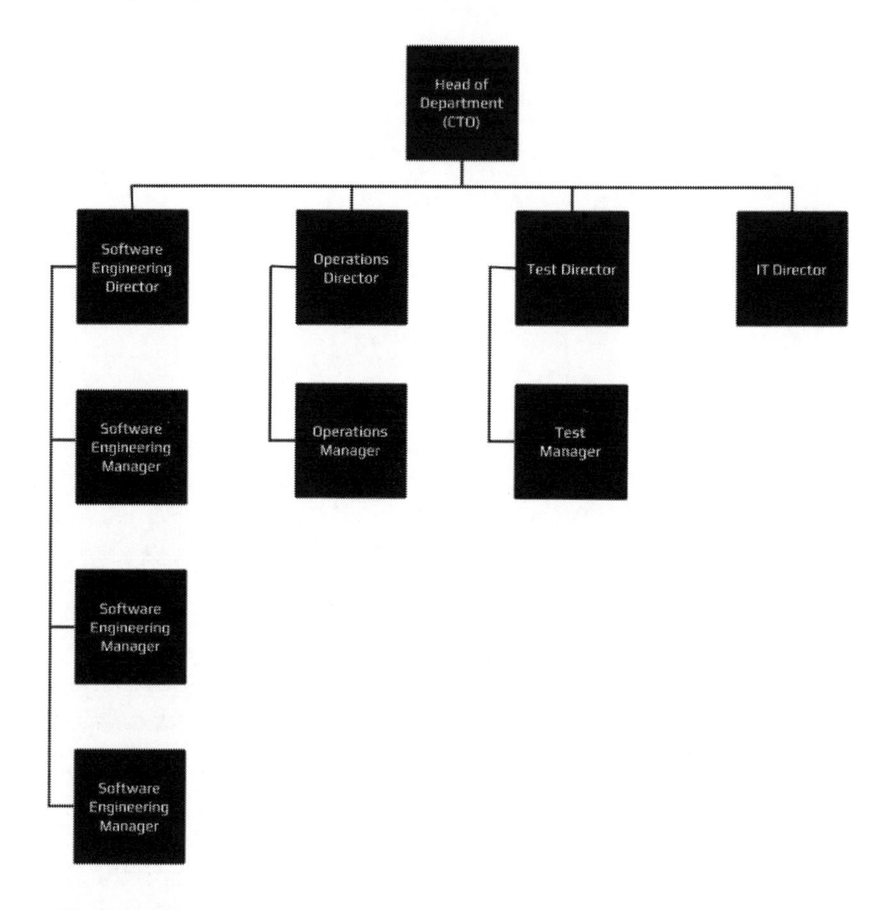

The functional organisation structure has its origins in the industrial revolution. At this time, factory owners saw the benefit of specialist workers, particularly in terms of training, and so grouped those performing the same tasks together.

The first recorded formal implementation of the functional organisation was at Swift & Co in the early 1900s. A wholesale butcher in Chicago, Gustavus Swift, organised Swift & Co into functional divisions. This company is credited by Ford founder Henry Ford as his inspiration for the production line.

Functional organisations are best suited to the production of standardised goods, at large volumes and for low cost. Those residing within this type of structure are characterised as efficient, but they struggle to work with other teams. Communication between teams tends to be formal in nature, leading functional organisations to be slower and less flexible than firms organised along other lines.

The teams within traditionally structured IT organisations certainly struggle to work together and, in fact, often oppose each other. IT organisations don't create standardised goods in large volumes, but they do require speed and flexibility. Why would they, therefore, choose a functional organisational model?

IT leaders, in the misguided view that departmental structures should reflect the structure of the business, often choose this model. But it is commonly observed during organisational restructuring that structures often fail because they are over-applied. So a firm's structure doesn't necessarily provide a good model for that of individual departments within the business.

IT leaders might have chosen a functional organisation structure because of habit. There's a growing body of evidence that suggests that when behaviours are repeated habitually, they require no conscious thought. If leaders have experienced nothing but functional organisations, they might recreate those structures without being conscious that they're making a choice at all.

A third possibility is cultural lethargy. It's only really in the last 10 years that IT has been producing online services almost exclusively. Prior to that, much of the IT industry was still producing packaged software products. A functional organisation suits the production of software far more than the provision of online services - although it's still far from ideal for that.

Functional vs Divisional and the product-centric IT organisation

Whether a conscious decision or otherwise, an organisational structure is a choice - and this has a profound effect on employees.

So when an organisation needs its teams to collaborate and creates structures that impede this very aim, employees receive a confused message. When people are confused, they become introspective and focus on what they're most comfortable with.

It's no wonder then that software, operations and test engineers focus on their disciplines and become more specialist in favour of stepping outside of their boundaries and working together. The functional organisation was designed to achieve this after all.

This, if nothing else, demonstrates the power of the organisation structure: even without any conscious effort from leadership, it encourages the people working within it to behave in certain ways.

An alternative to the functional organisational structure is the 'divisional structure'. This typically splits large businesses into smaller, product-focussed organisations. It's common for banks to have commercial, retail and credit divisions for example. During my last years at AOL, the company organised itself into content, advertising and broadband divisions.

Divisional organisation structures have their roots in the post-World War II economic growth. At this time, the advent of computerised payroll, combined with the academic development of formalised management theories, allowed businesses to grow much larger than they had before. This in-turn led to a need for new organisational structures.

Some will argue that the divisional structure has disadvantages: for example, it is seen to encourage office politics and unhealthy rivalries. There is a concern that divisions will attempt to encroach on each other's territory and not share innovations. In my experience, however, politics only flourishes when the leadership thrives on it. Office politics can make no progress in an organisation that is focussed on achievement.

Groups within divisional organisational structures are autonomous, focussed on their products, and their achievements are measured by the success of these. The divisional model seems to offer the IT industry a far better alternative to the functional structure.

So why don't more IT organisations use divisional or product lines? Products are seen as transient and risky by many firms: they might not work, they could get dropped. Project requirements and, indeed, projects often change monthly, if not weekly. There is a fear that if teams are aligned too closely to individual projects or products, they will be demoralised by the increasing number of changes.

This assumes that product teams behave as functional teams do. Product teams, however, are actively engaged with their products and also play an active role in defining features and capabilities. As a result of this, they become actively engaged with any stakeholders outside of the IT organisation.

This further increases the chances of products' success and, more importantly, it gives teams some control over their destiny. They will participate in decisions that change the scope or the schedule and this limits the emotional impact of such changes.

There is also a concern that headcount will be removed from any team developing and supporting a product that is de-scoped. The solution is to align teams with multiple products; this ensures headcount numbers are assessed sensibly and applied efficiently.

The product-centric organisation creates teams aligned to own and take responsibility for the lifecycle of the products assigned to them.

They're all products!

The key to creating a product-focussed IT organisation is an environment where teams clearly own multiple products. If a team has several products in its portfolio and has complete creative control over at least some of these, a change to one solution has a much diminished impact on the team and its headcount.

Remember: the tools, systems and services that teams need to build their products are also products in their own right. All need roadmaps, development and integration support, as well as attention and focus, to a greater or lesser extent. This is the key to establishing teams engaged with the business of creating revenue-generating products, which also strive to improve the productivity and efficiency of the whole organisation.

A team building a revenue generating product might also own the source-control solution, artefact repository and deployment framework. This means that if the revenue generating product is struggling and its future becomes uncertain, the team can focus more on its other products.

Because the team still has essential solutions to develop, manage and support, it will feel a lot less exposed than if its only product's future was uncertain. This leaves the job of assessing how the product changes affect the future of the organisation to the leadership – and the team can continue work on the products it provides to its peers. As result, all the other teams benefit – as does the IT organisation as a whole.

The trick is get the product set right. This is a decision that can only be made by the IT organisation's leadership, which should use its knowledge of the business, market and technologies to work with direct reports and create an environment where teams can focus on the products that are important to them.

Teams could service client organisations: for example, a team might be focussed on the marketing or retail products. Alternatively, teams could be aligned to particular technologies, with one owning all the Java or Python products.

Teams with a product requiring significant development work might also own the solutions needed by that process, such as the source control and automated build systems and artefact repository. However teams are focussed, the most important thing is that this is stated clearly as part of the IT strategy.

One thing I can guarantee is, everyone will have an opinion on the products they work with every day. Teams will naturally engage with each other over those products and that intense shared interest will break down any barriers that might exist between them. Anyone who has any doubt over this should ask a developer what they think of the source control system they use at work - or question an operations engineer on their deployment mechanism.

IT strategy and the Org structure

The strategy and structure of the IT organisation communicates powerful messages to the people within it as well as to the rest of the business. It's therefore very important that both should be crafted to reinforce the message, rather than confuse it.

If the organisation is looking to automate behavioural tests, the IT strategy must state that clearly and provide some guiding standards. The organisational structure should reinforce that by giving clear ownership of test tools and frameworks, providing implementing teams with clear routes for support and guidance. Formalised tactical approaches should be outlined to help teams share their experiences, as well as step definitions and tests, with others.

Further, leadership should provide opportunities to recognise those that have created innovative tests or optimisation mechanisms. This shouldn't be done through cheesy awards, but by allowing innovators to present their methods and discoveries to their peers, ideally both inside and outside of the business.

If the organisation is looking to implement a micro-service architecture, the IT strategy needs to provide guidance as to how these will be integrated. Given such an approach, it's incredibly important that everyone designing services is acutely aware of the standards they need to work to.

Again the strategy must make this clear. The organisational structure should provide teams with authority over their domains: a team must design, build and support its service. If the team that has authority over deployment, or performance is separate from those building the service, it completely destroys both parties' authority.

So teams must feel they own their service, but they must also understand the boundaries they work within. However, equally, they must know how to challenge these: this is another example of day-to-day implementation providing feedback into the tactical approach.

Finally, the business needs to ensure there are feedback loops between every level. Indeed, it would be rare for an organisation to get the strategy, tactical implementation and day-to-day coordination right on the first attempt. With appropriate feedback loops in place, there are opportunities to improve processes rapidly until an optimal approach is found.

Specialists and small teams

One question I'm always asked, is how this approach works for very small teams - or even individual specialists. Specialists can't necessarily be provided for each product team. Likewise, there may not be enough work for more. They could be matrixed in and work across all the product teams, but that would break the model and bring back old conflicts.

The solution lies in considering specialist capabilities as products, too. For example, take an organisation building multiple products that requires a specialist database administrator (DBA). There are two approaches that can be taken, both of which entail considering the specialist DBA capability as a product.

One approach is to give teams the freedom to contract specialist DBA experience as and when required. Teams could be encouraged to engage a consultancy, supported by a tactical approach that provides easy access to that service. Third-party consultancies have become adroit at growing and sharing specific domain knowledge among consultants.

Another approach is for the DBA to build this capability into a product within the IT organisation. Encouraging the DBA to provide products as a service - such as design consultancy, optimisation and troubleshooting - encourages groups to make use of these at the appropriate time.

By providing guidance detailing how and when particular products should be used, the DBA will waste less time trying to create solutions to problems that could have been avoided at the design stage.

This approach encourages the production of repeatable approaches and frameworks. Teams can use these to benefit from lessons learned during the development and management of other products. Because these would be DBA products, the onus would be on that specialism to formalise and share, rather than on a team more interested in satisfying business requirements.

Another benefit to this approach is, it becomes very obvious exactly how much that specialist capability is required. It is easy to justify additional capability if it's being exhausted, rather than driving an individual specialist too hard, forcing him or her to make under-the-counter compromises.

This same approach can be applied to architecture and network engineering - and the security benefits are even greater. Rather than trusting an individual specialist to review development across all products and the network, as well as any third-party interactions, teams can implement specific solutions provided by the security specialist.

Verifying the behaviour of well-defined products is a much better way of ensuring security than trusting it to the timely engagement of an individual expert when teams are on a deadline.

Transitioning to a product-centric organisation

Implementing any framework or model requires significant domain knowledge, intelligence and pragmatism. The exact approach will depend on many factors; I can't hope to cover them all. I can, however, provide some guidance as to how a transition should be approached.

The first step is to clearly understand and articulate the IT strategy. No change in organisational structure can possibly be successful unless it is enabled and supported by this.

Next, undertake a product review. This should list all the products the IT organisation builds as well as those it uses to do this. A one-line bash script that executes on a Cron to copy a file from one place to another is just as much a product as the source control system, or an e-commerce site, or an online game.

Once the review is complete, it will become very clear that the IT organisation uses many more products than was immediately obvious. The list will probably highlight multiple products with overlapping capability. It may be tempting to initiate a review of these, but this should be avoided at this stage. Such reviews rarely result in simple actions.

For now, the organisation should proceed, assuming all products are necessary: at the moment, they probably are, even if multiple products appear to be fulfilling the same function.

The next stage is to understand the skills and capabilities that are needed to build roadmaps, design, develop, launch, manage and support all the products on the list. This should be compiled in the form of a matrix in order to enable the next stage.

Once the matrix of products and skills is available, alignment strategies can be tested. I suggest that this is first done as a virtual exercise. Try creating teams focussing on particular technologies, for example, and see if there are any oddball technologies in use. Does that render a technology alignment inappropriate? Should those products be replaced by ones that sit better within the IT strategy? Try associating the products by function, such as development, testing or operations. Is that an appropriate? Does that reinforce or contradict the IT strategy?

Next, try focussing teams on particular customers or stakeholders. This approach can work well for an IT organisation that primarily builds products for internal customers. Again, test this by checking whether it contradicts or confuses the message of the IT strategy. Once the strategy has been chosen, the product alignments describe the portfolios that will be owned by each of the teams.

The product portfolios now need to be assessed to determine the composition of the teams that will own them. These products, all subject to lifecycles, will require different update frequencies, development, configuration and support activities.

It's easy to assume that a team owning mostly third-party provided products won't spend much time developing these. But this may, or may not, be the case. It could be that significant development work is required to integrate the products together.

Creating teams focussed on products is an activity that needs to be done with managers and leaders together. They will know their people and products best and understand how the various solutions are used. The new teams will be responsible for their products 24/7, so availability and temperament is as important as capability.

If the IT organisation wants to try out the approach without restructuring, it can choose one product portfolio and build a team for it. Because people must have the skills to manage the lifecycle of all products in the portfolio, it's likely that individuals will have to be pulled from existing teams to form this one.

The new team must be allowed time to find the optimal way to work together. If this is to be an experiment, the team must have a say in the criteria used to judge its success or failure. Everyone else in the IT organisation should be informed of the experiment and given details on what is hoped to be gained. When the experiment is finished, the results should be shared with everyone, ideally by the product team itself. Experiments aren't successful or unsuccessful. Hypothesis or theories are tested for validity, that's all.

If the organisation chooses to restructure itself entirely, the capabilities and temperaments of all staff should be mapped to product portfolios. At this point, it will probably seem there just aren't enough people. It will be very tempting to give up and determine that the method is flawed.

But while no method is perfect, this is not failure. It only appears that there aren't enough people because products weren't properly managed in the old structure: they were poorly designed, and not informed by the necessary requirements, with compromises made in secret. This doesn't mean an organisation hoping to become product-focused automatically needs to increase its headcount. This is where the IT strategy performs its coup-de-grace.

As I've stressed time and again, the IT strategy must explain not just what needs to be achieved, but how the goals will be met. I've made a point of specifying that the IT strategy needs to provide the standards that the organisation will work to. This is where such a strategy pays dividends.

Having realised there aren't enough people to fully staff every team and properly invest in every product, leadership now makes product decisions:

- Are there products that are good enough for now, that can remain static for a time?
- Are there in-house products requiring development effort that can be replaced by third party provided solutions?
- Are there products draining resources, which aren't actually providing a real benefit, that can be decommissioned?
- Are there standards that can be relaxed for certain products because they are less essential?
- Are there products which, when delivered, improve productivity and quality for the entire organisation and so should receive the lion's share of people and resources initially, but can then take a back seat?

The answers to these questions will form or change the tactical approach. This is a great early exposure to the way in which decisions will be made from this point forward. This balancing of headcount, resources and time is what gives the leadership of a product-focussed organisation explicit power and control over its activities and output.

The advantage of this method over the previous approach is that it allows the leadership of the IT organisation, fully aware of the consequences, to make the decisions about compromises itself. These aren't made de facto because no-one paid attention to something even though everyone else thought it had been handled. These aren't hasty decisions made because someone is shouting. This isn't negligence because a team is context switching and has missed something. These are product level decisions taken by a responsible and informed leadership.

Conclusion

No one would seriously suggest that modern IT organisations have anything in common with the textile mills or wholesale butchers of the industrial age. Yet they are following the same organisational structures pioneered by those industries.

Clearly these structures need to change if intelligent, self-motivated engineers are going to take responsibility for the development and management of internet-accessible services.

Next Gen DevOps provides a product-centric approach to analysing IT organisations, helping them transition to a modern structure that devolves authority and promotes autonomous, responsive and responsible teams.

CHAPTER NINE

The only successful DevOps model is product-centric

This chapter...

Will examine four innovative organisations that are the acknowledged leaders in their fields. We will see if their approaches share any common elements and if they do, how this can be modelled and replicated in other organisations.

The Unicorns

Netflix, Etsy, Facebook and Amazon have talked extensively about their journeys. They have presented at numerous conferences as well as appearing on multiple blogs and traditional print media.

These businesses are dubbed "The Unicorns" by commentators due to their apparent unique and special ability to work in ways that no other businesses can. This chapter will uncover the recipe for the special secret sauce used by these organisations - and show how this method can be made to work in any business.

Netflix

Adrian Cockroft, former Cloud Architect at Netflix, is open[16] about how engineers are organised and work at the company. The consistent theme is, the people developing software at Netflix are responsible for how it performs in the live environment. Everything else is about giving these people what they need to allow them to take that responsibility. The idea is to step back while they build the next generation of video content distribution.

At Netflix, people are organised into teams that allow them to build the products needed. An explicit understanding of Conway's Law has allowed Netflix - formally a micro-service based organisation - to organise itself into a micro-team organisation.

Netflix is a business with cross-functional teams capable of building, launching, managing and supporting all of its products. So teams working for the online streaming service have the tools, capability, responsibility and authority they need to build, deploy, monitor and support the new features the product needs.

Netflix was one of the first organisations to embrace Amazon's Cloud at scale. Yet the real innovation was building the tools that allowed the product teams to focus on building new features, ultimately providing a better service to customers. Netflix has always made a big deal out of these products and the list is now extensive.

---

[16] Adrian Cockroaft's infamous "NoOps" blog post: http://perfcap.blogspot.co.uk/2012/03/ops-devops-and-noops-at-netflix.html

Janitor Monkey looks for unused Amazon instances and terminates them; Suro is a data pipeline service for collecting, aggregating and dispatching events including log data; Scumblr searches websites for evidence that Netflix has been compromised; and Zuul is a service that provides dynamic routing and resiliency functions among other things (check out the Netflix' Github[17] repositories for the complete list).

These tools are what allows the product teams to build, launch, manage and support products at Netflix. These tools are operations at Netflix.

Etsy

Etsy is another organisation that has openly shared its journey to DevOps. Like Netflix, Etsy didn't start out following DevOps practices but has rapidly evolved how it's working practices over time.

Etsy's story - about collaboration and shared responsibility - sounds very human. Although they brought different skills and experiences to the table, the online retailer's engineers realised they had a lot in common with each other. In fact, during an interview[18] Michael Rembetsy, VP of Technical Operations at Etsy, compared the development of DevOps at the firm to a courtship, explaining that the relationship had to grow before marriage was possible.

---

17 https://github.com/Netflix

18 Network World Interview with Michael Rembetsy: http://www.networkworld.com/article/2886672/software/how-etsy-makes-devops-work.html

Etsy organises engineers into small teams with focussed missions. These are decided 60 days ahead of execution and are tracked and managed in two week increments.

Etsy has talked extensively about Deployinator[19] the tool it built and open sourced to handle its site deployments (recently relaunched as a Ruby Gem). However, a visit to Etsy's Github repositories[20] tells a similar story to Netflix: it builds tools for everything. Opsweekly is for on-call classification and alerting; StatsD for pulling real-time data from systems; Feature API for controlling how new features are rolled out to customers; and Nagios-Herald for adding context to alerting.

This is the story of operations' at Etsy multi-skilled engineers enabling the tools that allow them to take full responsibility for the features they build.

Facebook

Facebook's story is similar to Netflix and Etsy's but very different in scale. The social network also started out following a more traditional developer and ops split. It had operations silos that took responsibility for site reliability, but found this didn't allow them the responsiveness they felt they needed and embarked on a journey to evolve capabilities and capacity.

Facebook's Production Engineering department had been through several iterations before the social networking site hit on the model that works today. The only real difference between Netflix and Etsy is scale. Otherwise, all the firms organise engineers into small teams, giving them responsibility and authority over the features and systems they build. Facebook's meteoric growth

[19] https://github.com/etsy/deployinator
[20] https://github.com/etsy

provided additional nuance to the problems it faced but ultimately they were the same issues as its peers.

Facebook moves fast: in fact one of its mantras is, "move fast, break stuff". But in order to do this, the social network needed to devolve responsibility and authority to engineers tasked with building new capabilities. There just aren't enough amazing developers who are also amazing system and test engineers as well as being great network and database engineers - and data scientists. Therefore, Facebook needed to build systems and tools to support developers and give them the best systems, test, network and data capabilities.

In addition to these, there is also a clear thread of responsibility: all Facebook's software engineers participate in an on-call rotation and understand that the team that builds the features is responsible for them - with the operations team available to support and assist them.

The capabilities of Facebook's Production Engineering function can also be found on Github (are you starting to see a pattern emerge?). Grocery-delivery is a tool for syncing Chef (configuration management) objects; Bistro is for scheduling and running distributed tasks; Taste-tester is a test mechanism for Chef configuration management changes; and Watchman is a tool for watching files and firing triggers. Not available on Github but spoken of extensively are Gatekeeper and Airlock, tools Facebook uses to test and deploy its features. Gatekeeper and Airlock are so sophisticated, Facebook has even used them to prank the press.[21]

---

[21] http://apptimize.com/blog/2015/03/how-facebook-feature-flagged-its-way-into-a-feature-article/

Amazon

Amazon have ended up at the same place as Netflix, Etsy and Facebook and Amazon even managed to skip a step. The key to Amazon's ability to skip a step was a recognition of the power of APIs far earlier than anyone else.

I'll spare the gory details if you want to read the original source of this search for "Stevey's Google Platforms Rant". The long and short of it is that back in 2002 Jeff Bezos, CEO of Amazon, was so frustrated with how long it took for the various different teams at Amazon to build stakeholder engagement, agree terms and actually get thing done that he mandated that all teams must share their data through service interfaces. Teams must communicate with each other through these interfaces, there will be no other form of interprocess communication. The technology used doesn't matter just the use of programmatic interfaces. Further, and this Amazon have ended up at the same place as Netflix, Etsy and Facebook and Amazon even managed to skip a step. The key to Amazon's ability to skip a step was a recognition of the power of APIs far earlier than anyone else.

I'll spare the gory details if you want to read the original source of this search for "Stevey's Google Platforms Rant". The long and short of it is that back in 2002 Jeff Bezos, CEO of Amazon, was so frustrated with how long it took for the various different teams at Amazon to build stakeholder engagement, agree terms and actually get thing done that he mandated that all teams must share their data through service interfaces. Teams must communicate with each other through these interfaces, there will be no other form of interprocess communication. The technology used doesn't matter just the use of programmatic interfaces. Further, and this is the magic one, all interfaces must be designed to be exposed externally.

If you work from first principles and follow the thread of these mandates you can draw a straight line from designing the technology to run one of the biggest web retailers in the world all the way to being able to sell Cloud services.

There's no Github repository for Amazon's tools because they sell them. They are Amazon Web Services. If you want to know how Amazon builds servers start an EC2 instance. If you want to know how Amazon manages server configuration and orchestration look at AMIs and Cloudformation, if you want to know how Amazon monitors it's systems check out Cloudwatch.

API's were the key to Amazon being able to skip the step of building employee collaboration and relationships. There is still plenty of collaboration at Amazon but it's born from the capability AWS provides not from the necessity of building stakeholder support.is the magic one, all interfaces must be designed to be exposed externally.

If you work from first principles and follow the thread of these mandates you can draw a straight line from designing the technology to run one of the biggest web retailers in the world all the way to being able to sell Cloud services.

There's no Github repository for Amazon's tools because they sell them. They are Amazon Web Services. If you want to know how Amazon builds servers start an EC2 instance. If you want to know how Amazon manages server configuration and orchestration look at AMIs and Cloudformation, if you want to know how Amazon monitors it's systems check out Cloudwatch.

API's were the key to Amazon being able to skip the step of building employee collaboration and relationships. There is still plenty of collaboration at Amazon but it's born from the capability AWS provides not from the necessity of building stakeholder support.

Special secret sauce recipe

It can be seen that all of these organisations took different routes to DevOps, but the components of their transformation are almost identical:

Deployment, infrastructure provision, configuration management, monitoring and security are products that must be made available to product teams.

There must be nothing that blocks progress or stops teams taking responsibility for every aspect of their work. If something is blocking progress, authority is required to engineer around this. If progress is hindered by process or people, the teams must be free to engineer solutions to these problems. If there is a common issue shared by several teams, new products are needed to resolve it.

Teams must be small enough to work well together.

Five to seven appears to be a sweet spot for management, collaboration and innovation.

Products that support building, launching, managing and supporting revenue generating products should be constructed with APIs.

Further, these APIs should be constructed with the idea of being shared externally. This ensures new hires can be immediately productive and encourages standard approaches that can be easily shared between teams.

Operate in public.

Code should be open source and shared on Github. Obviously this excludes the organisation's core intellectual property.

If an organisation follows this recipe, it can gain a bunch of additional benefits for free. Engineers will take responsibility for their work; productivity and innovation will not slow as the organisation grows. Engineers deliver their best work because their code is available for public review. Additionally, they gain public recognition - a great motivator which makes finding new engineers much easier: potential hires know that they'll also have this opportunity. Finally, publishing open source software makes a cast iron case for research and development tax breaks.

This openness is taken a step further too, all of these organisations have shared their journey with the world, they operate in public, and their failures as well as their successes are subject for debate.

Conclusion

Netflix has a little over 2,000 employees, Etsy around 750, Facebook 11,000 and Amazon 154,000. Playfish had around 140 employees when I left, Connected Homes had about 50 while I was there. So scale is irrelevant: these techniques work in businesses with 50 to 150,000 employees.

None of these organisations started out as DevOps focussed. They've all evolved and iterated on their approach, arriving at DevOps. They've come to the position via different routes but with the understanding that they need products to help them build products. By my definition, at least, they are product-centric organisations.

Amazon was able to skip a step because Jeff Bezos realised the power of programmatic interfaces. This enabled the biggest web retailer in the world to launch and remain the biggest cloud provider in the world.

There are no unicorns here.

Instead, there's a realisation that monitoring, alerting, configuration management, deployment, fault tolerance, security and testing are products which need to be engineered.

Now, thanks to all of these companies contributions organisations don't have build all of these tools themselves. It's possible to create a Github account and download all this code. In addition to this, the Next Gen DevOps Transformation Framework (also available on Github)[22] provides a roadmap that organisations can use to plan their own DevOps journey.

---

[22] https://github.com/grjsmith/NGDO-Transformation-Framework

CHAPTER TEN

What does a 21st Century IT organisation look like?

In this chapter…

We'll see what it's like to work for - and lead - a product-focussed IT organisation like the one we proposed in the previous chapter. This chapter will examine what the teams look like and how they behave, describing how they approach common tasks such as sprint planning, development, deployment and incident management.

We'll look at the mindset of the people in the team and the changes they'll have been through, showing the compromises they have had to make.

Finally, we'll consider some guiding principles for a product-focussed IT organisation.

Introduction to the product team

Superficially, the product team will appear like any other. Its morning will start with coffee and a discussion over any incidents that occurred the previous night.

But looking a little closer will reveal this team, and all of those around it, glancing at dashboards while jackets come off and backpacks are put down. Minds are unconsciously matching patterns, checking for anomalies. The dashboards they are glancing at display a mixture of business, application and system metrics, as well as some custom metrics highlighted as problem indicators.

What won't be obvious to the casual observer is that the project team is comprised of software, operations and test engineers. When each is ready to start work, he or she will glance at the stories on the wall, observing which are taken, in progress, closed, and which remain open.

Each team member will each run a ticketing system to review the comments on the ticket last worked on: any of these could be a story describing a product behaviour or an incident report. Each engineer's specific field only becomes apparent if one pays attention to the tickets themselves, or tracks the comments on them.

The operations engineers could be coding new infrastructure configuration and creating a deployment and update mechanism. The software engineers might be investigating an incident ticket raised by the alerting system. Software and test engineers could be working hand-in-hand to create a test for a new feature.

All engineers are working with source-control, automated build systems and test harnesses. At the same time, all are reviewing their products' dashboards throughout the day while commenting on tickets, coding new capabilities and investigating product behaviour.

Management

The team's manager keeps track of progress, using the ticketing system while reviewing reports and burn-down charts from the previous sprint. The manager will also be representing the team's interest in other products that fall into other portfolios. It may be that this team requires a new step definition that's not present in the test harness. The manager will be discussing this with other teams' managers, finding out if anyone else is investigating it and if there is any interest in collaborating on the work.

The line manager will check calendars and respond to holiday requests, attend one-to-one meetings with the engineers and perform the usual day-to-day task management. The difference is that the tasks are not confined to a single discipline. The product team manager is in charge of product tasks, whether they are related to operations, software or test engineering.

The product team manager has one significant additional task that the line managers of single discipline teams don't have: he or she will keep track of product performance. After all, the manager's performance, as well as that of the team, is judged, in part by the success of the products.

While not a product manager, the line manager is now an active part of product management. The role will entail looking for patterns in product performance and actively seeking out the next initiatives needed by the team to improve on this. The line manager will work with product sponsors, managers and any other stakeholders to prioritise the stories addressed by the team. This manager brings the unique value of having intimate familiarity with the technologies involved in the product. Product team managers are ideally placed to find the interesting information that predicts performance.

When Playfish released Simcity Social, the studio's tech lead decided that he wanted to maintain a ratio of Java exceptions to player numbers. If the number of Java exceptions recorded increased significantly without a corresponding rise in player numbers, he wanted it to be investigated. This was just one of his contributions, but it was the one that demonstrated that he was an integral part of managing that game as a service.

Leadership

The product-focussed IT organisation's leadership assesses the performance of the revenue generating products against strategic objectives. This entails preparing product decisions to guide teams to hit those targets. If a product is under-performing, leadership must work with other functions within the business to create an action plan. If the resolution can be delivered by the IT organisation, the new activity will need to be expressed in terms of behaviour. Either a product requires a new feature, or higher quality, or performance needs to be improved.

Once the desired behaviour has been decided, leadership can work with the product team manager to create an action plan. This may be as simple as the creation and addition of a few stories into the next sprint, or it might be much more complex.

It could be that, in order for performance to be optimised further, whole new systems are needed. The product's performance problem might fall to an authentication mechanism that uses a simple database lookup. The solution could entail moving to a key value store or to add a caching mechanism. Either of these can require a whole new product.

Leadership then needs to decide which team has the skills and capacity to create this new product. If no such team exists, it must decide how the new product will be provided. For example, perhaps new feature development can be slowed on some other products for a time while the new solution is built.

Leadership will have to work with the sponsors of revenue-generating products as well as team managers to ensure they fully understand the ramifications of such a decision.

Products can be incredibly dependant on each other, as this specific example shows. The traditional solution of bringing in a third party to provide a product is still possible. However, ownership of the product will have to be assigned eventually: any team that takes this on is going to suffer more significant problems in the long term when compared with a team that provided the solution itself.

Leadership, sponsors, management and engineers all work together to make these product decisions, which in turn allow the organisation to make progress as a whole.

The same sort of process occurs at budget time, when product decisions are made after a reassessment.

If budget reviews necessitate major changes to the product portfolios, leadership will work with the affected teams to ensure that they are still viable. If a team loses one product of a portfolio of five, it could present an opportunity to devote more time to another - or some products could change ownership.

While products shouldn't be reshuffled lightly, this is a reality: teams will need to have their expectations set accordingly.

How do the team feel?

Every member of the team will have to make adjustments and accept some compromises. Software and test engineers must get used to being on-call: the whole team needs to take responsibility for its products. The engineer on-call will have clear escalation paths and access to support from their other team members. This ensures that problems can be investigated and mitigated quickly and with minimal stress.

Over the weeks and months, cross-training will take place, with every person in the team able to check for the most common problems and undertake rudimentary troubleshooting steps. Of course, repeat incidents are rare, because they are tracked down and resolved rapidly: after all, they affect the whole team.

Operations engineers will have to learn how to actually work in a team. In traditional operations, there is very little actual teamwork: most activities are considered solo.

In the past, only one person was needed to work on a ticket, log on to a server, update configuration or search log files. But that isn't how operations works anymore and the profession is still learning to adjust.

Updating configuration in a modern operations team entails writing and testing code, creating and testing builds, planning and executing roll-outs. These are most definitely team activities and operations is still getting to grips with what that means. Operations engineers in a product team will lose the flexibility they're used to. They'll feel scrutinised, given that they are accustomed to working with very little supervision. Now they will have a scrum master who needs regular updates - and the team is waiting for their work to be finished, so that other stories can progress.

I saw some of this at Playfish, when I removed some of the operations engineers from the ticket queue and put them in a project team to develop the configuration management system. Significant development work was required to extend Chef to meet the requirements of the games teams and EA's central operations function.

I hired a scrum master to manage the project and under his guidance, created some epics to get the team started. At first, the team relished the opportunity, but it soon became clear that the pressure of writing code to a deadline had taken its toll. People began to feel this formality was unnecessary: tempers frayed, progress stuttered. But over just a couple of sprints, it became business-as-usual: progress improved and the team was proud of what it had achieved.

The benefits of this team approach result in a much smoother progression through the various test stages. The team works together to agree how code should be included and activated later, so that builds can be deployed with incomplete features included.

Because the entire team is intimately familiar with the design of features, the test process, the tests that are executed at each stage and the risks inherent in each build, there is never a block to progression at any stage. If new environments are needed, they are provided within and by the team, in a timely fashion, because there's no barrier to communication between the software and operations functions.

This means there's less frustration: individual teams aren't forced to 'hurry up and wait'. Environments aren't suddenly required at the last minute. Testing isn't delayed while testers are made available or take time to prepare scripts. Software engineers aren't frustrated by an ill-prepared operations team asking seemingly irrelevant questions about the design of new features.

There is a general air of progress and cautious optimism about the team. There's no fear of the unknown. To the product team, problems that weren't foreseen don't lead to blame and finger pointing, but to interest and inquiry.

After all, the build has been tested by software, operations and test engineers working together. A battery of ever-increasing, automated tests have been executed against the build on at least two environments with different combinations of stubs, mocks and test and live services.

Any new problem that occurs in live is a challenge worthy of investigation and the team takes pride in coding tests to ensure the issue is taken into account in the future. Any problems found during the development process or during the testing of a build are dealt with when they come up, not at the end of the sprint, or at deploy time.

This is a team building a product together which will share and learn from with each and improve together.

How do Ops, Dev and Testing interact?

The expectations of others have a huge impact on how people behave - and on their performance. Software, operations and test engineers tend to come into their professions from different backgrounds and perceive the world in subtly different ways. However, the differences aren't as great as you may think.

They behave differently at work because of the structures they operate within, and the contrasting expectations that people have. Given an opportunity to take responsibility and a commensurate level of authority, engineers will thrive, whatever their discipline and background.

Software engineers tend towards optimism: it's in the nature of the profession to assume any problem can be overcome. Test engineers tend to be conservative but curious. Operations engineers tend towards conservatism, if not outright pessimism; they are also driven to strive for efficiency and are personally offended when it is lacking.

This set of traits forms an excellent basis for planning. Presented with new requirements, software engineers imagine designs and relish the challenge. Test engineers work out how to verify the behaviour, considering how they'll explore around the new feature. At the same time, operations engineers will consider how the new feature will affect performance, questioning whether any new monitoring is required and if any changes to the environments is needed. When combined, these behaviours strike a balance between optimism and caution when the team is called upon to provide its estimate.

Over time, team members will behave less like individuals from different disciplines, instead learning how to get the best from each other. Planning will be a team discussion and this will provide a consensus, rather than individual responses.

At first, the team will split up into individual disciplines as features are built. This will see software engineers code new features; test engineers code tests; and operations engineers code infrastructure changes. They'll come together only at milestone events, such as when the first formal test runs take place.

But over time, the team will begin to discover the value that comes from working together. This will start with design tasks: software and operations engineers will pair when designing high performance features. Software and test engineers will partner to develop and create tests for complex features.

As the engineers will learn, many features benefit from pairing with other disciplines. This creates a 'best of both worlds' scenario: engineers retain their individual specialisations, at the same time reaping the rewards of working closely with other disciplines.

The product team is not a group of generalists: organisations don't need to hunt for the semi-mythical full-stack developer. The product team is an environment that makes the best use of specialists and encourages them to work together and learn from each other.

Most of us in IT have seen this happen occasionally, generally when there's a crisis. When Playfish launched the Sims Social, the war-room comprised operations, software and test engineers, as well as community managers and business intelligence people. This cross-functional team scaled the entire Playfish infrastructure and addressed major weaknesses in the platform services in just a few short weeks.

Tiger teams have been commonplace since the 1960s, but still we only use them in crisis situations. It's been said that the scrum team is the tiger team without a crisis, but many businesses still populate the their scrum teams with just software engineers, perhaps adding a tester or two. Imagine what might be possible if every team was a tiger team, focussed on product success everyday.

What do the environments look like?

Working methods aren't the only change seen with the adoption of product teams. The environments used for development, testing and even live environments will transform significantly too.

Most IT organisations fail to use infrastructure efficiently. Everyone is aware of this, but no one knows what to do about it.

Often, far more infrastructure is used than necessary because projects aren't initially set up with testing in mind. Test environments exactly mirror live ones when it's unnecessary, because the teams don't understand the nuances of operating system, hypervisor or network behaviour.

Infrastructure choices are made by people with insufficient knowledge or experience. Products continue running long after they are needed often with no clear ownership, and no one willing to make the decision to decommission them.

Testers coming late to a project have to rely so much on trust that they are never quite sure whether issues are being caused by poorly set up environments, or real software problems. This in turn forces software engineers to duplicate the entire test case again in the development environment before they can start looking for the bug.

This amalgamates into more work, time and servers than necessary – and all because IT organisations consider the engineering disciplines individually and chronologically. Software engineers are hired first, followed by testers and then operations. After all, there's no point having a tester when there's nothing to test yet, or operations people when the product isn't ready for launch – is there?

The product team approaches things differently. It starts with the necessary disciplines in place, enabling it to make sensible decisions from the outset. This team wants to ensure the product is easy to develop, test and manage in the live environment. People are focussed on the fact that products are services that will run for years and that they will have to perform thousands of deployments over this time. The team will consider that before there's a product, there needs to be a test harness ready to execute the first tests. Before there's a build, there must be an appropriate environment built and then hosted.

Some of the earliest advantages the multi-discipline product team brings are to the build environments. The team's focus will be on ensuring builds are created quickly and automated tests executed as rapidly as possible. This presents all kinds of design opportunities.

If the build is dependant on many third-party libraries, it may be appropriate to download these into a local repository before it begins. If the build is built from many distinct elements, a machine containing SSDs might provide distinct advantages. If the build is, in fact, multiple software artefacts, a multi-processor solution could improve performance.

System integration benefits more than most from the attentions of a multi-discipline product team. As the application matures and acquires interfaces, decisions are needed as to whether they will be stubbed, mocked, connected to a test service, or the full live solution.

Each of these decisions is far better made by a team aware not just of the function of a service, but its likely performance. Often, there are so many of these decisions that by the time the application is close to functionally complete, it appears like a black box to anyone not involved.

This often forces test and operations engineers to attempt to deduce behaviour, which takes considerably longer than if they were an active part of the design process and helping to make each decision.

A detailed knowledge of the exact function and use case for each interface also allows tests to be automated more easily. It encourages operations engineers to make decisions that could considerably speed up the creation and testing of the integration build.

When a service is trusted, has provenance and the build only requires a few different responses, it can be safely mocked. This sort of decision can look very suspicious to test and operations engineers who come late to the product. But it is completely benign and eminently sensible if taken early on by a multi-discipline product team fully aware of the implications. It also might make load testing considerably easier in the future.

In situations where services are still under very active development and not yet well trusted, the team is faced with deciding whether to test against the live version, or - if the service is also being developed in-house - a latest working build or release candidate version.

Early appreciation of this might lead the team to build flexibility into the integration process to allow it to be tested against any - or all - of these different versions of the service. This is an area where traditionally, a lot of infrastructure is wasted as teams demand multiple copies of the entire live environment so they can test against many versions of just one service.

If the team has the right mix of skills at the start of the project, it may be possible to create all these different environments on a single piece of infrastructure, and even perform tests against multiple versions of the same service simultaneously.

Live environments benefit too. From the most simplistic viewpoint, software engineers aren't necessarily the right people to be making infrastructure and hosting decisions. However, there are far greater opportunities than that available to the product team.

A product team focussed on building a service to be responsible for once it has launched will be in a good position to ensure monitoring is added in the right places. This reduces the need for deductions using circumstantial evidence. And reducing the number of tests the monitoring system has to execute has a huge impact on its performance, scalability and efficiency.

Log files are the closest thing to a real-time view of the function and performance of the product. Log design, formatting and logging levels can make a system transparent and provide teams with useful real-time data, which can be used to make decisions. But they can also cloud the real activity of a system and render a team blind to the performance of a system.

Mindful of this, the product team must build monitoring and testing throughout the development process. The team can optimise logging to get the most value from it, which can save a fortune in terms of services or development time to process logs.

A few weeks after the launch of the Sims Social, Playfish's game servers were logging two gigabytes (GB) of log data every second across the application servers. There might have been useful data in those, but we'll never know.

There was a move, later on, to use one of the many logging services to help process the data. During the trial, a mistake was made in the build: the application logged so much data that it breached its thresholds. The project never progressed any further.

This was completely unnecessary, but it is what happens when logs are considered as an afterthought. That much data had never been logged in the development environment. Then again, the development environment had never seen 125,000 simultaneous users.

Failures are a fact of life. Hardware fails; upstream configurations cause problems; products experience unexpected behaviour that they weren't designed to cope with. Making products deal with failure without losing revenue or customers is extremely difficult, time consuming and expensive once the solution has been built.

However, it doesn't have to be this way: combine the natural pessimism and experience of operations people with the talent and ingenuity of software engineers and the tenacity and curiosity of testers and products become much more tolerant of failure. Success comes naturally as the product is built.

During the planning phase, as designs start to take shape, they are proposed, discussed, analysed and refined by the team as a whole. When risky choices are made, mitigation can be taken at the time, rather than bolted on later. As a result, products become increasingly tolerant of failure and also more efficient: they are designed not just for functionality, but also for performance and reliability, from day one.

The guiding principles of the product-focussed IT organisation

Product first

The IT organisation's first concern must be its products - and this also applies to the product team. All the initial choices and decisions should be derived from product requirements. Whether the product needs 24/7 support, zero downtime, high performance over 3G, or instant access should influence the composition, focus and behaviour of the team required to build and manage it.

Tests

These days, everything the product team builds is a complex system - and it is connected to other complex systems. The only sensible response to such an environment is to verify behaviour. The product team lives by its tests. Anything that can't be proven systematically is suspect and requires thorough manual exploration, this is time consuming and hence costly so the team must endeavour to minimise it.

Skills

To the product team, skills and capability matter: roles and job titles don't. The product team measures its success by that of its products and is motivated by this.

Conclusion

Consider major developments in IT over the last 10 years. Open source has replaced commercial off-the-shelf software. Agile development has solved the problem of large, out of control projects that were unresponsive to changing markets.

Test engineering and release management emerged. Cloud computing removed the need for significant initial capital investment. Online services became ubiquitous.

At the same time, billion dollar companies shifted away from traditional models and became online businesses. E-commerce evolved from being about shops, to centring around ecosystems. DevOps grew from a base of a few conferences focussed on tools and configuration management and began to be discussed as an operating model.

Yet throughout all of this, IT organisations are persisting in their old habits. They try and adopt new technologies and techniques, but only manage to make small, limited gains. Larger organisations continue to treat engineers as if they are low-skilled and lazy.

Everyone is frustrated. Everyone is plagued by a feeling that they could do so much more.

In a modern IT organisation, teams shouldn't have to negotiate with each other to build, launch and manage products. Structures shouldn't exist that encourage teams to consider their contribution in any way divorced from the product. Testing shouldn't be an afterthought supplied by the lowest bidder.

Major IT developments over the last 10 years have created an urgent need for change. The pace of that change is accelerating. Next Gen DevOps presents a framework for modern IT organisations to build and manage successful online services and keep pace with change. By adopting these ideas, processes and methods, firms can be sure they are fully equipped to take on the new IT world and adapt to changes and take advantage of new opportunities.

CHAPTER ELEVEN

The Next Gen DevOps Transformation Framework

This chapter...

Will look at The Next Gen DevOps Transformation Framework, outlining how to use it in practice.

To get the full benefit of this chapter, review the framework on Github: https://github.com/grjsmith/NGDO-Transformation-Framework

What is the Next Gen DevOps Transformation Framework

From the README: The Next Gen DevOps Transformation Framework is an attempt to provide organisations with a structured approach to a DevOps transition.

The framework is presented so that anyone reading it can understand the nature and complexity of a transition to DevOps. It also allows organisations to choose which challenge to tackle first, with each project structured to provide clear benefits. In addition, the framework demonstrates further functions that can be enjoyed when combining the capabilities delivered by multiple projects.

The framework is presented as a series of tables to ensure the reader understands the flow from one capability level to the next. While this makes editing the individual markdown documents less convenient, it was a necessary compromise to ensure the project's goals were met.

When to use the framework

The framework has been designed to be as practical as possible and to support any conceivable journey to DevOps. If an organisation is undertaking a complete restructure, the framework can be used to structure the programme. For an individual team looking to improve its capabilities, the framework will define the projects likely to have the most impact.

Because the framework is presented as a series of projects, it's also possible to use it in reverse: Firms can identify the capability required and determine a series of projects that will lead to it.

How does the framework support a DevOps Transformation?

Inspired by ITIL's maturity model, the Next Gen DevOps Transformation Framework is structured as a series of capability models.

Each capability collects together a sequence of functions that modern organisations need in order to build, launch, manage and support online services throughout their lifecycles. The capabilities (in alphabetical order) are:

3rd Party Component Management
Budgets
Build & Integration
Change Management
Feedback Loops
Incident Management
Project Management
Service Engineering
Test Engineering

These capabilities aren't team names or job titles: the size and nature of an organisation will dictate where each function exists. The capabilities are presented with a series of levels, describing increasingly capable approaches to that function. For example:

Build & Integration Level 0 is named 'Build & Integration: Ad hoc'. It is described as: "Continuous build ad hoc and sporadically successful."

An organisation whose Build & Integration capabilities match the Next Gen DevOps Framework definition of Level 0 Build & Integration will recognise its experience in the Observed Behaviours section:

• Only revenue-generating applications are subject to automated build and test.

• Automated builds fail frequently.

• No clear ownership of build system capability, reliability and performance.

• No clear ownership of 3rd-party components.

If an organisation recognises its behaviour in these descriptions, it can undertake the projects described in the Project Scope associated with Level 0. In this case, two projects must be completed in order to reach Level 1:

• Create a process for integrating new 3rd party components.
• Give one group clear ownership and budget for the performance, capability and reliability of the build system.

By formalising how groups integrate 3rd party components such as libraries or device drivers into their software, builds become more reliable. This might be achieved by storing the particular versions used in a build in a local repository, or by ensuring the repository these are pulled from maintains old versions.

Giving one team clear ownership and budget for the performance, capability and reliability of the build system is the first step in recognising the build & integration process as a product in its own right. Unmanaged build systems quickly slow down, either because the infrastructure scaling doesn't match the usage, or because plug-ins are added and never actively managed - so the build system requires an increasing amount of resources that might not be necessary to build the software.

These projects are classic examples of enablers. They provide value in their own right by improving build reliability and performance - as well as putting additional capabilities within reach. If a group has ownership of the build system, it's a lot easier to move towards continuous build, integration and eventually delivery, because the implementation of those capabilities can be planned, costed and people assigned to make it happen.

Due to the information's complex nature, the framework doesn’t display well on an ereader or a small format paperback book. This extract from the Build & Integration section is enough to show the increasing capability available to an organisation using the framework.

Capability Level	Capability name	Description	Observed behaviour	Project scope
0	Build & Integration: Ad hoc	Continuous build ad hoc and sporadically successful.	* Only revenue generating applications are subject to automated build and test. * Automated builds fail frequently. * No clear ownership of build system capability, reliability and performance. * No clear ownership of 3rd-party components.	* Create a process for integrating new 3rd party components. * Give one group clear ownership and budget for the performance, capability and reliability of the build system.
1	Build & Integration: Continuous Build	Continuous build is reliable for revenue generating applications.	* Automated build and test activities are reliable in the build environment. * Deployment of applications to production environment is unreliable. * Software and test engineers concerned that system configuration is the to blame.	* Add environment tests to automated test system to be executed in the build environment, add similar tests to the monitoring service to validate essential configuration elements match. * Resolve all failing tests with Configuration Management system ensuring all configuration artefacts (config files, reference data etc.) are in source control. * Add system build and configuration tasks to build system. * Ensure all engineers understand why configuration elements vary between systems and environments.
2	Build & Integration: Continuous deployment	Continuous deployment of individual services is reliable for all products, services and tools.	* Configuration for systems, virtual machines and containers can be deployed without human intervention * A minimum level of automated testing is in place to confirm applications deployed and display minimal viable functionality * Service changes can only be deployed individually and core functionality changes require large, dedicated cross-functional project teams	* Add dependency management and discovery to configuration management and deployment systems * Create mocks for 3rd party dependencies that don't provide test mechanisms * Add more sophisticated automated tests to assure core service functions operate as expected

This model is repeated to enable other capabilities. Let's look at another example:

Feedback Loops Level 0 is named: 'Feedback Loops: System Metrics' and is described: "Service function and performance is inferred from measurement of system metrics."

An organisation whose Feedback Loop capabilities match the Next Gen DevOps Framework definition of Level 0 Feedback Loops will recognise the behaviours described in the Observed Behaviours column:

• Services are monitored at system and container level, not from within the service itself.
• Services log errors to files, which are at best partially parsed, checking only for known errors.
• Service failures are commonly reported by users before they are captured by the monitoring system.

Organisations that recognise these behaviours can undertake the projects described in the Project Scope associated with Level 0. In this case, two projects need to be completed in order to reach Level 1:

• Create custom metrics that describe service health by combining available business and technical metrics.
• Analyse all available data from logs, database entries and system events, looking for correlations that confirm actual behaviour of the service and its users. Monitor this data and iterate alerting thresholds.

Combining system and business metrics to create custom service metrics makes the monitoring data meaningful to everyone in the business, as well as taking the guesswork out of service performance. 60% CPU utilisation doesn't provide any clue as to whether the service is performing properly. However, the fact that it takes three seconds to complete a billing transaction gives everyone in the business a real understanding of service health.

Correlating anecdotal experience of service performance with system and business metrics allows the creation of thresholds that remove all doubt about performance. It then becomes possible to alert when performance becomes unacceptable prior to complete service failure, significantly reducing the impact of incidents.

Here's an extract from the Feedback Loops table taken from the framework. As with the Build & Integration table, capability increases as projects are completed.

Capability Level	Capability name	Description	Observed behaviour	Project scope
0	Feedback Loops: System Metrics	Service function and performance is inferred from measurement of system metrics	* Services are monitored at the system and container level not from within the service itself. * Services log errors to files which are at best partially parsed checking only for known errors. * Service failures are commonly reported by users before they are captured by the monitoring system.	* Create custom metrics that describe service health by combining available business and technical metrics. * Analyse all available data from logs, database entries and system events looking for correlations that confirm actual behaviour of the service and it's users. Monitor this data and iterate alerting thresholds.
1	Feedback Loops: Correlated Metrics	Core service functionality is monitored indirectly.	* A variety of business and system metrics from multiple sources offer some assurance of service function. * More service failures are captured by the monitoring system than the users but incident triage takes too long.	* Invest development time in capturing key service events and their performance in real-time. Record these events and the performance data in an easily searchable object store. Aggregate the data and add suitable thresholds in the alerting system.
2	Feedback Loops: Service & System	Core service functionality and performance is monitored directly.	The service itself records functionality checkpoints and performance metrics in a high-performance store making simple correlations and presentation of data trivial. Senior leadership and non-operations people still feel abstracted from the service and user experience.	* Correlate service data with obfuscated customer and transaction data and present it on dashboards. * Build more sophisticated monitoring and alerting thresholds based on this new customer centric data such as alerting by geography, ISP, or CRM data.

These two capabilities were not chosen at random. They are an example of a combination of projects that once completed, put additional capabilities within reach.

Once the build system is under consistent and organised management and it's possible to measure custom metrics that describe the actual performance of the service, only a little additional work is required to start monitoring how long the build process takes. These metrics can help identify when the build system needs additional capacity.

Once the Level 0 Test Engineering projects are completed, it becomes possible to monitor automated performance tests in the build environment. These results can be used to predict how new features and fixes will affect the performance of the service in the live environment.

These additional capabilities are illustrated with the cloud icon on the capability web diagram included with the framework. These are by no means intended to be exhaustive; they are just some of the more obvious incidental benefits available once multiple projects have been completed.

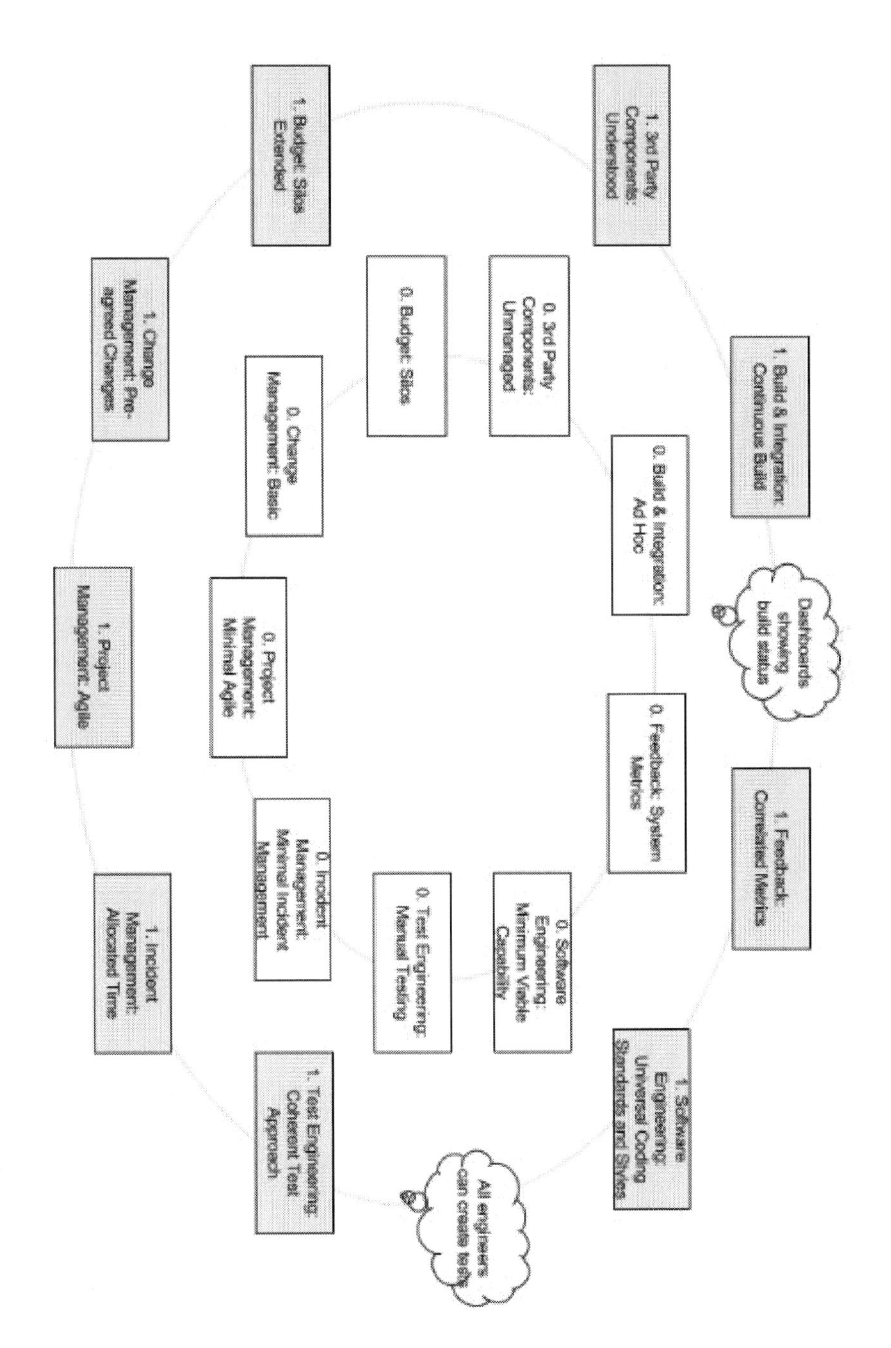

see exhibited and the capability level that maps to those. For example, if an organisation wants to observe the following behaviours:

- Automated tests exist for all user stories and use cases.

• Manual checks are no longer required to ensure core service functionality.

• Test engineers focus their time on exploratory testing and improving the sophistication of the test systems.

These are the observed behaviours for Test Engineering Level 3.

In order to achieve Level 3, an organisation needs to have:

• Formalised and published a coherent test approach reviewed by all engineers.

• Chosen an automated test mechanism and framework that supports a human readable domain specific language (DSL).

• Coded the steps necessary to allow simple user interaction functionality (such as form submission and check status type actions) to be tested at build time.

• Provided training to testers enabling them to write automated using the DSL tests.

• Had test engineers create user stories or use cases for the steps needed to automate more of the standard tests.

• Involved test engineers in use case or user story creation to ensure new feature use cases or user stories have appropriate test steps specified or tests created.

• Developed test steps and tests alongside new features.

• Focussed test engineers on UX and ensured all major user activities are assessed at build time using automated tests.

This could form the basis of a requirements document for the types of organisations that work in this way.

Conclusion

However large or small an organisation, however teams are structured, the Next Gen DevOps Transformation Framework offers a series of practical steps to achieve the rapid response IT needed to ensure success in the modern world of online services.

The projects described in the framework have been designed to require collaboration between software, operations and test engineers and will build trust between those groups as they are completed. High trust between teams is one of the indicators of a high performing IT organisation.[23]

[23] Having a high-trust culture has a strong impact on both IT performance and organizational performance.
http://www.thoughtworks.com/insights/blog/state-devops

CHAPTER TWELVE

Appendix to chapter 2: The role of women in the history of operations

This essay...

Will show some of the more surprising aspects of the technology industry's history.

Hundreds of articles have been written about early computers, designers, engineers and even programmers. However, information about the operators is less available. Why? The programmers of the first computer systems were also the operators.

I talk about this in detail in the second chapter of my book, The History of Operations. But beyond this, I made some surprising discoveries about the role women played in the history of computer operations.

Pioneering men AND women

Much has been written about Admiral Grace Hopper, but it is not well known that she was more than a programming pioneer; she was also an operations pioneer.

In 20 years in the technology industry, I've only worked with two female system engineers, but I've worked alongside many women developers. So, while it was surprising to find a woman on the very first computer team, I wasn't shocked. But digging a little deeper into the Harvard Mark 1 team, it is interesting to see that the programmers were supported by female mathematicians. They would help prepare the equations that Hopper, Bloch and Campbell would programme the Harvard Mark I.

A few years down the line, the ENIAC attracted many female programmers. Some of these women went further and modified the machine, adding the capability to store programmes. But their contribution was hidden from the public and photos doctored to conceal their identities.

This version of an often reprinted U.S. Army Photo shows Cpl. Irwin Goldstein as he sets the switches on one of the ENIAC's function tables at the Moore School of Electrical Engineering, courtesy Harold Breaux.

In this U.S. Army Photo Cpl. Irwin Goldstein can be seen and now so can some of his female colleagues.

The picture of the team that built EDSAC (Electronic Delay Storage Automatic Calculator) - the UK's first computer and the precursor to the world's first commercial computer LEO, at Cambridge University - is even more striking.

As the picture shows, many of the people who built the UK's first computer were women. If the technology industry was

[24] University of Cambridge Mathematical Laboratory members May 1949

Top row, from left: D.Willis, J.Stanley, L.Foreman, G.Stevens, S.Barton, P.Farmer, P.Chamberlain

Middle row, from left: H.Smith, C.Mumford, H.Pye, A.Thomas, E.McKee, J.Steel

Bottom row, from left: R.Bonham-Carter, E.Mutch, W.Renwick, M.Wilkes, J.Bennett, D.Wheeler, B.Worsley founded on significant contributions by both men and women, why are there so few women working in IT now?

The Computer Girls

In a bid to answer this question, I began to research women in technology in the 1960s. This lead me to two men who would open up a whole new world of fascinating insights.

Nathan Ensmenger, an associate professor in the School of Informatics and Computing at Indiana University and author of "The Computer Boys Take Over" specialises in gender and computing. His work led me to a Cosmopolitan Magazine article: "The Computer Girls."

The Computer Girls

A trainee gets $8,000 a year
...a girl "senior systems analyst"
gets $20,000—and up!
Maybe it's time to investigate....

Written in 1967, it's a patronising piece encouraging women to seek a career in technology. The article even quotes Admiral Hopper comparing programming to planning a dinner party.

Ensmenger has uncovered many other fascinating stories. For example, Bobbi Johnson, the Miss USA 1964 winner who fulfilled her ambition of becoming a programmer and went on to work for General Electric as an application engineer.

The other person to open the door to the world of computer operators during the 1960s was Larry Luckham, who managed a data centre at Bell Labs akin to the set from The Bionic Man.

Photo copyright Larry Luckham, Reproduced by permission.[25]

During the late 1960s, Luckham worked alongside more female computer operators than I've ever met.

[25] http://www.luckham.org/LHL.Bell%20Labs%20Days.html

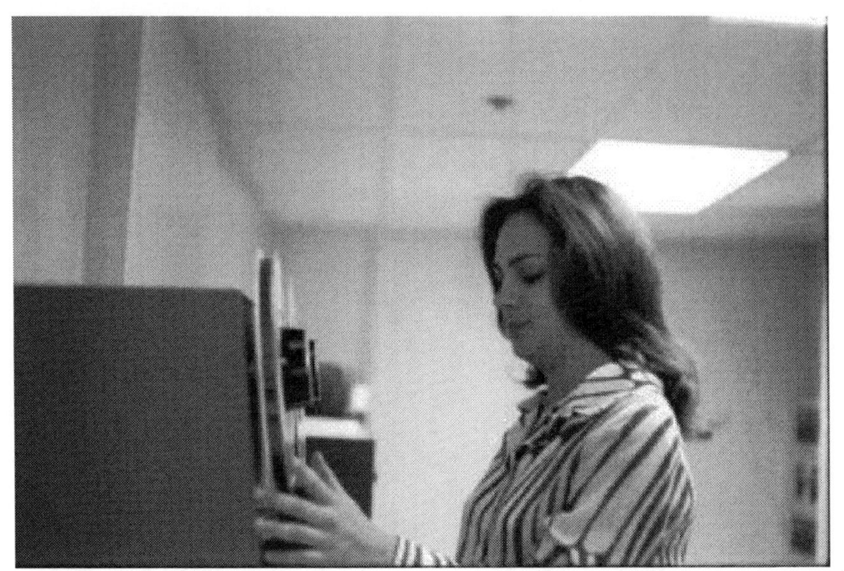

Photos copyright Larry Luckham, Reproduced by permission.[26]

I spoke to Luckham about his data centre. He explained: "The Bell Labs data centre that I ran had an IMB 360/50 mainframe with 512 K of ferrite core memory, a card reader/punch, a chain printer, five nine track tape drives, and three Winchester disk drive arrays. Each of those mounted nine removable hard disks that held 29 MB of data. Only eight of the nine could be online at any given time; one was a spare. So, there was 696 MB of online data."

Luckham's site is packed full of pictures of his team. The only man in the photos other than Luckham is a programmer. All the computer operators seem to be women.

167% growth!

The 1970s brought minicomputers as well as the Unix operating system. The major industries were now computerised and there was a huge push for these machines to start delivering the promised

[26] http://www.luckham.org/LHL.Bell%20Labs%20Days.html

benefits. It led to standardised methods for programming and testing as well as for managing programmers and operators. Alongside improvements in process isolation and system resiliency, these methods served to increase the effectiveness of computing systems. The minicomputers then pushed computing from head offices to regional and allowed end users to interact directly with the machines. Computing capabilities expanded from data analysis and calculation to data recording and booking systems.

With the increased capabilities of computers came a corresponding need for talent. The number of people working in computing increased dramatically. It's difficult to know exact figures: many jobs required computers but weren't necessarily recorded as computing professions. A good indicator is the number of people that graduated with Computer Science degrees:

"The percentage of women graduating with a Bachelor's degree in computer science rose by 167% from 1970 to 1983 (i.e., from 13.6% to 36.3%). (Camp 1997)[27] This initial upward trend in the 1970's corresponds to dramatic increases in the numbers of computer science graduates in this newly formed field (from 2,388 in 1970-71 to 24,510 in 1982-83). "[28]

---

[27] Camp, T. (1997) "The incrediblc shrinking pipeline", Communications of the ACM, 40(10), 103-110. Paper is available on-line at at: http://womendev.acm.org/archives/documents/finalreport.pdf

[28] Women in Computer Science: Where Have We Been and Where are We Going? Tracy Camp, The Colorado School of Mines, Golden, CO Denise Gurer, 3Com Corporation, Santa Clara, CA: http://citeseerx.ist.psu.edu/viewdoc/download?doi=10.1.1.127.5603&rep=rep1&type=pdf

Brochure for the PDP-11 resource timesharing system, the RSTS-11 system, and BASIC-PLUS software.

So it can be postulated that there was something like a tenfold increase in people working at an expert level with computers during the 1970s. This doesn't seem unreasonable given that… "it's estimated that by the time the PDP-8 ceased production in 1979, around 300,000 had been sold.[29]"

[29] The Digital Equipment Corporation PDP-8

So as the capabilities of computers grew the need for skilled people in the Technology industry grew.

What happened in the '80s stays in the '80s

The 1980s was a time of great technology investment. The US and UK were moving away from primary industries, seeing huge investment in the finance and service sectors.

Networking, UNIX workstations and PCs saw computers more widely distributed: they were finding their way on to desks both at work and at home.

Yet something else was happening: "from 1984 to 1994, the percentage of women graduating with a Bachelor's degree in computer science decreased by 23.4% (from 37.1% to 28.4%). "[30]

Was this the turning point; the reason I've rarely met any female system engineers? I decided to take a look at the data in more detail.[31]

Frequently Asked Questions
Part of the PDP-8 Collection by Douglas W. Jones THE UNIVERSITY OF IOWA Department of Computer Science: http://homepage.cs.uiowa.edu/~jones/pdp8/faqs/

[30] Women in Computer Science: Where Have We Been and Where are We Going?
Tracy Camp, The Colorado School of Mines, Golden, CO
Denise Gurer, 3Com Corporation, Santa Clara, CA
http://citeseerx.ist.psu.edu/viewdoc/download?doi=10.1.1.127.5603&rep=rep1&type=pdf

[31] http://www.nsf.gov/statistics/nsf13327/content.cfm?pub_id=4266&id=2

I started with Computer Science.

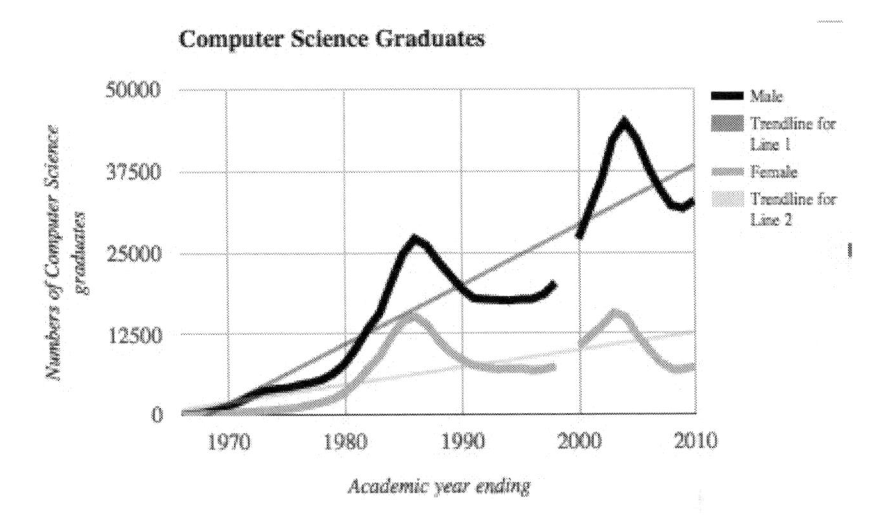

The trend lines make it very clear that while Computer Science courses are attracting more females year-on-year, women are taking up the degree at a slower rate than men. Interestingly, the pronounced drop in women graduating with Computer Science degrees is echoed by their male counterparts. Whatever happened in the mid-80's affected both men and women.

Let's look at the results from some other courses, starting with my chosen field of study, Civil Engineering. I chose Civil Engineering because it's a multi-discipline course: We studied maths, physics and geology as well as practical courses including surveying, design and project management. Like Computer Science, it attempts to cram multiple different skills into one course.

When I started my degree in 1992, there were over 100 people in my first class, six of whom were women. I always found it interesting that while we lost a lot of people during the length of the course (hours were long, the work was tough and the professors unsympathetic) all the women finished it.

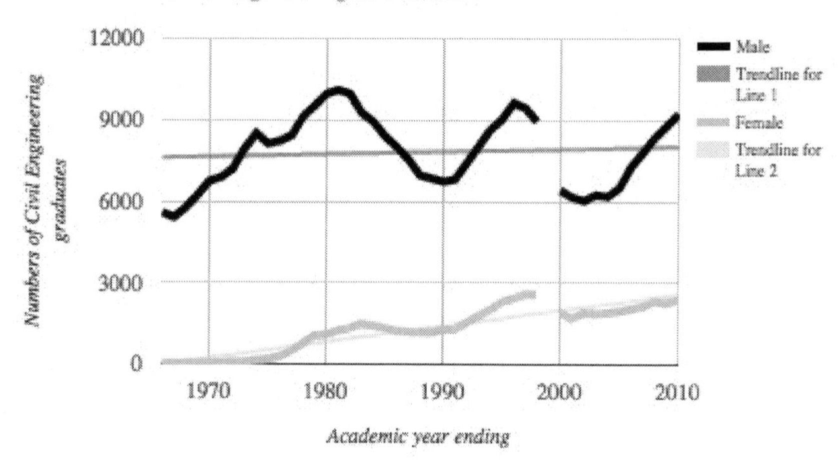

Here we see the same dip in the mid-80s - although it starts a little earlier. One thing that stands out is the variation in the number of men graduating compared to women. The rate of women graduating appears to be accelerating faster.

Maths was the favourite field of study for the early computing pioneers.

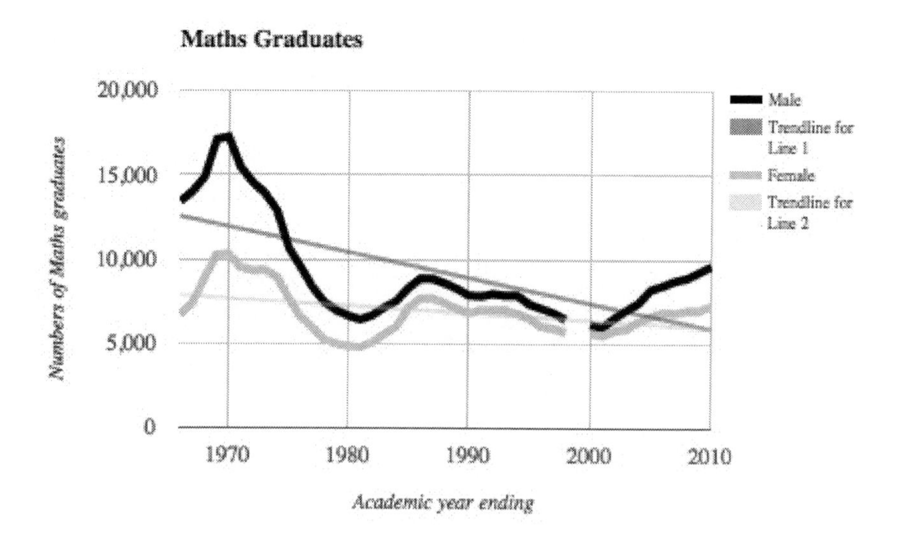

While Computer Science and Civil Engineering graduate numbers were growing in the mid-70s, Maths was shrinking. There was a slight pick-up in the mid-80s but the subject is either less popular, or it's getting a lot harder to graduate with a Maths degree.

While I was studying Civil Engineering, my girlfriend at the time was studying Physics. She always complained that there weren't many women choosing Physics and the National Science Foundation's statistics certainly echo her experience.

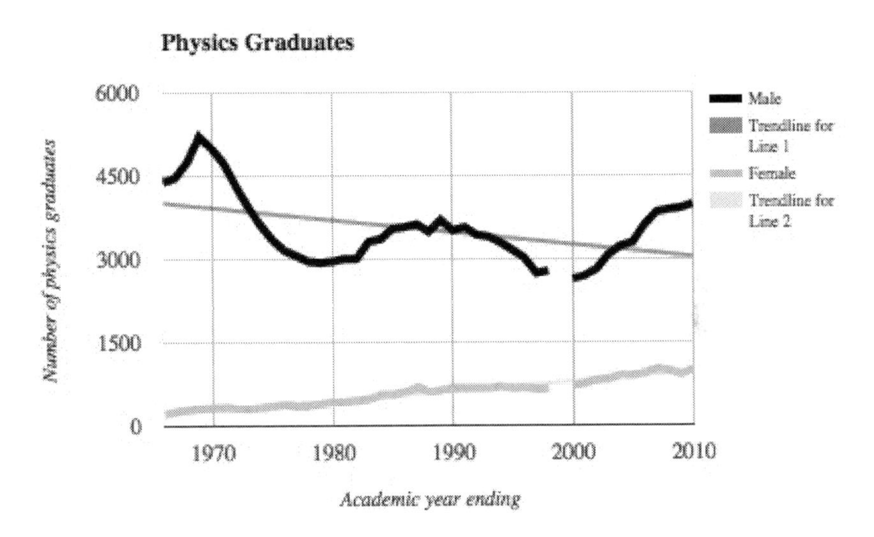

An increasing number of women are graduating with Physics degrees. However, less men are completing the course.

So far we're seeing a growing number of women each year graduating with Science, Technology and Engineering degrees. Maths is slightly less popular.

Moving away from the harder science and engineering fields, there are courses might appeal to a different type of person.

Take economics, for example.

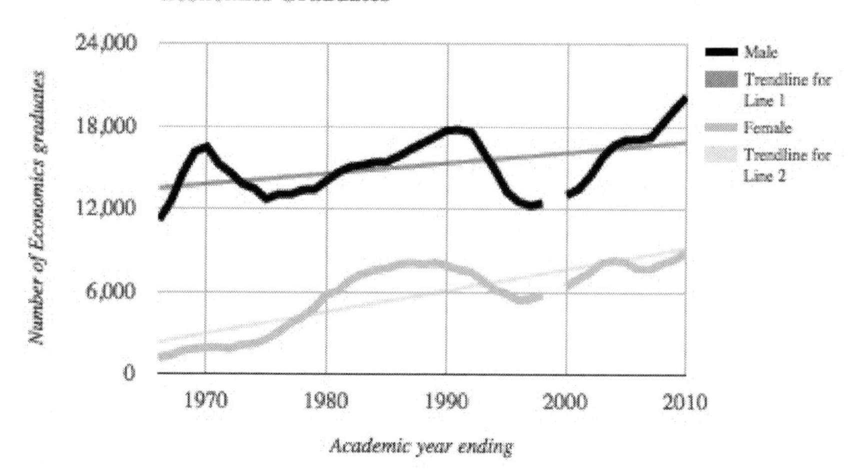

Economics appears to be more appealing to women than men year-on-year. While most courses experienced a dip in terms of graduates in the mid-80s, economics shows a steady increase or even a slight bump.

Let's take a look at one last course: Political Science.

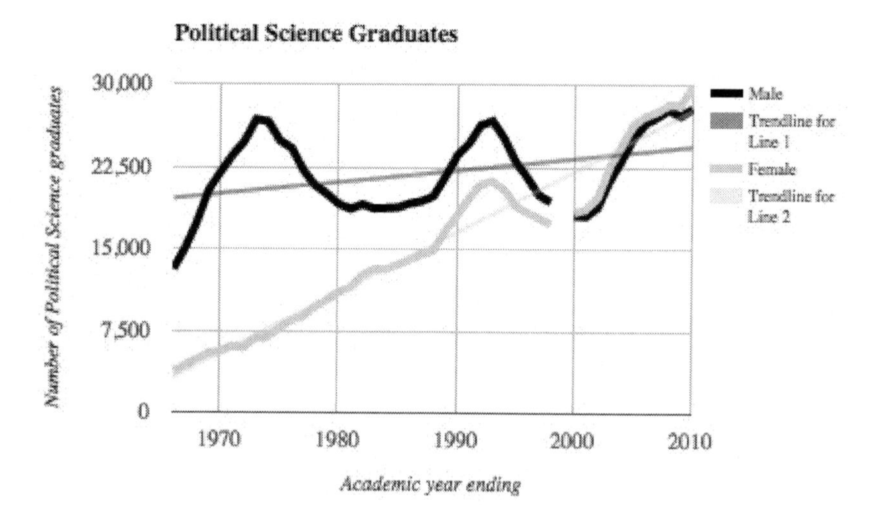

While most of the Science, Technology and Engineering courses I've examined have seen increasing numbers of women graduates, none of them have come close to the growth rates seen in Political Science.

Conclusion

It's clear that Computer Science degrees are increasingly popular with both men and women. However, while a growing number of women are graduating with Computer Science degrees, the gender gap is widening because men are graduating at a faster rate.

The drop in the rate of women graduating with Computer Science degrees appears to have an impact on other science, technology and engineering courses too.

Let's consider a hypothesis. As all the course graduate numbers change at similar times, the variance could well just be due to population numbers. So it doesn't look like anything happened to women graduating with Computer Science degrees in the mid-80s that didn't also happen to men.

But what about the pure numbers? If you didn't notice the y-axis of all those graphs, go back and look again. Only Economics and Politics have anything similar to the numbers of Computer Science graduates - and the latter is almost as popular as both of the former combined. While we'd all like to see more talented people entering the technology industry of either sex, we already have many more people graduating with Computer Science qualifications than other science, technology and engineering degrees.

If we dismiss the variance and look at the trend lines, Computer Science is increasingly popular with men and women, particularly compared to other courses. It's just a lot more popular with men.

My view is: I prefer to work in diverse environments. I think men behave differently when women are part of the workforce: the environment becomes more tolerant and collaborative and that results in better products and increased productivity. I know I'm not alone in that view.

But I don't think we need to make Computer Science more appealing to women. I think we need more degree courses.

Where should our future security experts, network, operations and test engineers learn the basics of their trade? Where should they learn about source control, or about contributing to a mature code base as part of a team of engineers?

Security, operations, network and software engineering are large enough skill-sets that they would benefit from their own degrees, each with relevant design patterns, test modules and practicals.

Finally, I think that focusing on the collaborative and social aspect of building, launching, managing and supporting modern online services during degree courses would encourage a group of people who might not otherwise see a future for themselves in the technology industry, especially if all they've seen so far is the Computer Science syllabus.

Printed in Great Britain
by Amazon